Knowledge Development in Transnational Projects

MW00718869

Transnational learning has become a buzz phrase in European policy-making and in multi-national business. Learning from the experiences of others is an idea that captivates practitioners and academics alike due to its simplicity and availability in a world that is increasingly characterised by cross-border and global connections. European regions in particular offer a diverse range of solutions to often shared challenges. This provides a knowledge base for other regions to draw on, through regional success stories, publications of 'best practice' and EU cooperation programmes.

This book explores 'transnational learning and knowledge transfer' in cooperation programmes and projects. It argues that a deeper understanding of learning needs to be central to the implementation of programmes and projects in order to successfully meet their desired outcomes and goals. By characterising some of the most important preconditions of transnational learning and introducing a process perspective to learning and transfer, this book identifies barriers to learning and knowledge transfer and contributes to a stronger conceptualisation of the topic. In doing so, it opens up the 'black-box' of transnational learning and knowledge development, providing a better understanding of its inner mechanisms. It also provides practical recommendations for policy makers and practitioners involved both at the programme and project level of transnational EU initiatives. This book will be of interest to students, researchers, and policy makers alike working in geography, political studies, legal studies and European studies.

Dr. Verena Hachmann has over ten years of experience with transnational cooperation as a researcher, and as coordinator and evaluator of both projects and programmes. After having been involved in EU and Nordic cooperation, she currently works as programme coordinator for a research and cooperation programme in the transport field at the Norwegian Research Council.

Knowledge Development in Transnational Projects

Verena Hachmann

Routledge
Taylor & Francis Group

LONDON AND NEW YORK

First published 2016 by Routledge

2 Park Square, Milton Park, Abingdon, Oxfordshire OX14 4RN
52 Vanderbilt Avenue, New York, NY 10017

Routledge is an imprint of the Taylor & Francis Group, an informa business

First issued in paperback 2020

British Library Cataloguing in Publication Data
A catalogue record for this book is available from the British Library

Library of Congress Cataloging in Publication Data
Names: Hachmann, Verena, author.
Title: Knowledge development in transnational projects / by Verena
   Hachmann.
Description: Burlington, VT : Ashgate, 2016. | Includes bibliographical
   references and index.
Identifiers: LCCN 2015039285| ISBN 9781472455833
   (hardback : alk. paper) | Subjects: LCSH: Regional planning—European
   Union countries. | Regionalism—European Union countries. | European
   cooperation. | INTERREG (Initiative) | Transnationalism—Political
   aspects.
Classification: LCC HT395.E8 H34 2016 | DDC 307.1/2—dc23
LC record available at http://lccn.loc.gov/2015039285

ISBN: 978-1-4724-5583-3 (hbk)
ISBN: 978-0-367-66835-8 (pbk)

Typeset in Times New Roman
by Swales & Willis Ltd, Exeter, Devon, UK

Contents

Figures

Tables

Abbreviations

BBSR	Federal Institute for Research on Building, Urban Affairs and Spatial Development, Bonn, Germany
DG Regio	Directorate General Regional Policy
ERDF	European Regional Development Fund
ESDP	European Spatial Development Perspective
ESF	European Social Fund
ESPON	European Spatial Planning Observation Network
ETC	European Territorial Cooperation
EU	European Union
INTERREG	Former Community Initiative of the ERDF, now European Territorial Cooperation objective of the ERDF
LCA	life cycle analysis
MCDA	multi-criteria decision analysis
NGO	non-governmental organisation
NWE	Northwest Europe programme area in INTERREG B
OECD	Organisation for Economic Co-operation and Development
PSC	Programme Steering Committee
SECI	'Socialisation, Externalisation, Combination, Internalisation' model developed by Nonaka and Takeuchi (1995)
SME	small- and medium-sized enterprises
UK	United Kingdom
URBACT	European programme promoting sustainable urban development

1 Knowledge Development and Learning in Transnational Cooperation

An Introduction

Learning from the experiences of others – the idea captivates by its simplicity and availability in a world that is increasingly characterised by cross-border and even global connections between people, organisations and regions. In recent years, 'learning' has featured strongly in scientific and policy discussions and publications – in various disciplines and often with a European perspective. Similarly, 'transnational learning' has become a buzzword in European policy-making, as in multi-national business. Its proliferation may be seen, for example, in the EU funding programmes that allow actors from different countries to get to know and make use of what is already tried and tested by their peers, or in calls for joint development across borders.

In this context, the diversity of European regions and their approaches to similar challenges is understood as a valuable source of knowledge and experience that can be tapped into through proceses of transnational learning – allowing an efficient spread of new ideas and solutions and enabling more regions to capitalise on innovation and opportunities for economic success.

In the EU context, transnational learning is mainly stimulated through a myriad of cooperation projects funded in thematic programmes, such as those for research and innovation or regional and urban development. Seen across the various EU funding programmes, considerable amounts are invested in transnational cooperation projects, and expectations of achieving verifiable results have increased throughout the years. However, despite the increased attention to learning aspects, there is not yet much knowledge about how learning manifests itself in the practice of cooperation projects, and little is known about the inner functioning of transnational cooperation projects. This knowledge would seem of particular importance in light of findings from past programme evaluations that point to modest project achievements and the inability to fully live up to programme expectations – for example, projects that do not manage to achieve all projected objectives and accomplish more moderate impacts than expected (see Panteia et al. 2010; Wink 2010; Dühr and Nadin 2007; Böhme et al. 2003). Overall, there are strong indications that many cooperation projects do not fully use the potential that lies in their transnationality, interdisciplinarity and practical orientation, while the causes for deficiencies in project achievement remain largely underexplored. Nevertheless, relevant research so far mainly concentrates on assessing the results and impacts of cooperation projects and programmes in various thematic fields. This leaves a considerable potential to further develop the knowledge dimension of transnational funding programmes by making use of the lessons learned both at programme and project level as building blocks for the future.

INTERREG as an Illustration of Transnational Cooperation

Of all the EU's funding programmes, its INTERREG programmes are probably the most dedicated to transnational cooperation, where one of its funding strands exclusively applies transnational cooperation as the modus operandi of its projects (further information on the INTERREG programmes can be found in Chapter 2). This is why INTERREG is used as an illustration of practical transnational cooperation for the purpose of this book. These programmes support projects in the field of regional development with a focus on innovation, transport, energy and resource efficiency and practical implementation, although many findings are transferable to similar transnational initiatives.

The Key to Producing Project Results: Transnational Learning

In transnational cooperation, achieving project objectives depends on many factors and processes, including project management, (inter-cultural) communication and personal motivation, but particularly on relevant knowledge development and learning processes. Transnational projects are funded to share and develop knowledge that forms the basis for producing the envisaged results. Project results are expected to be of use to all participating partners, but usually also beyond the immediate partnership, and to be transferable to other settings and regions. As this transferability of project results is an important part of the raison d'être of transnational cooperation, knowledge gains and lessons learned are vital project results. Thus, the knowledge dimension of transnational projects is also promoted by requirements for their result transferability.

In the transnational strand of the INTERREG programmes (strand B), the knowledge and learning dimension of transnational cooperation is furthermore heightened by their demands for achieving actual change and thus their very applied orientation. Projects are expected to go beyond research and analysis and embrace the implementation of new findings. They should either lay the ground for change or actually induce change. As achieving projected change is usually linked to human reflection, consideration and decision-making, it is a process of knowledge acquisition and expansion, in other words of learning. Knowledge-based project results strongly depend on the preceding knowledge production. Even very hands-on projects that aim at implementing practical solutions are closely linked to knowledge development processes at their basis. In the broadest sense, it could thus be argued that the objectives of transnational cooperation in INTERREG are achieved through knowledge-creation.

Projects of the transnational INTERREG strand are particularly interesting to study as they are characterised by distinct features that bring a rather complex perspective to knowledge development and learning:

- The projects' *transnationality* is multi-layered and not only relates to the partnership, but also to the whole project strategy: this includes a transnational objective, distinct and common challenges to be tackled and a joint working approach.
- Unlike permanent or long-term networks, transnational INTERREG projects are *time-limited projects* with a specific objective. Their knowledge development process is highly influenced by both the time limitation and their project strategy.
- Projects are characterised by a strong *practical and implementation-oriented focus* that goes beyond general exchange and networking. They bring together European actors working on common tasks related to regional development in a transnational consortium. It could be argued that this work-sharing approach increases projects' transnationality while some other EU programmes lack the joint action approach (for example interregional INTERREG projects are more characterised by a general exchange).
- Other transnational programmes also include pronounced processes of knowledge creation (for example Horizon 2020), but the large number of practitioners involved affects the distinct character of knowledge creation in transnational INTERREG projects. This makes knowledge development and learning more diverse than in cooperation programmes characterised by one dominant actor type (such as researchers).

Hence, the specificity of these projects lies in the linking of a transnational cooperation process with the implementation process of a distinct objective in a limited time frame.

With the new INTERREG V generation, demands have increased for more result-oriented projects and programmes. Each new funding period introduced new and advanced requirements for projects in order to better equip them to fulfil their expectations. As a reaction to an identified lack of impact, the 2000–2006 funding period saw claims for 'tangible projects' to strengthen the projects' implementation-orientation. Before, many projects had been conducted as rather open-ended networks, where the most important achievements were network building and a summary of 'best practices'. As a reaction to many projects focusing on local action, the 2007–2012 funding period called for more 'strategic projects' and truly 'transnational structures'. This development, making the expectations towards programmes and projects ever more explicitly ambitious, calls for a better understanding of what key challenges these projects face, and of how project processes could be structured in order to increase their chances of success.

The Book's Objectives: Conceptualising Transnational Knowledge Development and Learning

This book aims at including a process perspective on projects as well as increased awareness for knowledge and learning in transnational cooperation, thereby contributing to the advancement of research on transnational cooperation. To further develop this field, it focuses on transnational knowledge development and learning processes relevant to a variety of EU programmes and beyond, the challenges faced by cooperation projects, and the impact of these challenges on projects' ability to produce joint results. The book proposes a theoretical conceptualisation of transnational learning, a framework that is still lacking despite the wide application of transnational cooperation as the operational mode of EU projects.

It targets researchers in the field of regional studies, geography, international comparative studies, public management and organisational studies. At the same time, its target group goes beyond the research sphere, reaching out to policy makers and transnational project managers interested in enhancing the quality of results of projects, programmes and other forms of transnational cooperation. It provides programme stakeholders including a variety of programme institutions, the European Commission and national representatives involved in programme design and management, with insights into the internal functionality of projects and an environment conducive to the effective and efficient execution of these projects. The book furthermore helps project developers and managers in a broad range of EU-funded programmes and beyond to better understand the inner mechanisms of the specific project type they are dealing with in order to achieve project objectives in a more effective way. The better target-oriented projects function, and the more they produce and apply innovative knowledge, the better they can contribute to more target-oriented and effective programmes and ultimately to higher achievements with respect to their overall policy objectives.

This book combines theoretical approaches from research on organisational and inter-organisational learning and research on policy transfer to provide a framework for the analysis of the preconditions of transnational learning and the potential interactions of relevant project elements. The developed theoretical model is applied to practical transnational cooperation projects by way of deconstructing these into their structures and processes. This approach allows opening up the 'black box' of transnational learning and knowledge development to understand some of the relevant inner mechanisms of transnational projects. Critical challenges to transnational cooperation processes and the causes that may prevent projects from fully making use of their potential and mission are identified. This, in turn, provides a ground for practical recommendations to project and programming stakeholders beyond general project management advice. Linked to this, a better comprehension of the process of acquiring new knowledge through transnational cooperation also supports the evaluation of learning and its effects, an aspect that is often missed in the assessment of EU programmes, which is still dominated by quantitative aspects.

Specifically, the book pursues the following two objectives:

1 *To better understand the influence of project structures on learning processes.*
   This can support projects in designing useful cooperation strategies and helps
   to actively reflect on factors supportive to the achievement of planned results.
   Increased comprehension can also assist programme actors in advising on projects and in formulating appropriate requirements and standards. Moreover, it
   helps in selecting the most promising projects for funding.
2 *To better understand the reciprocal relationship of projects' learning processes and the results they deliver* (see Figure 1.1). This understanding can
   help to better grasp the challenges of transnational cooperation and how programme and project stakeholders can overcome these.

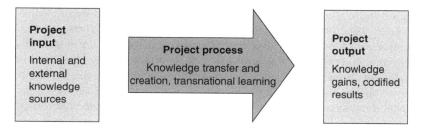

Figure 1.1 General project process

Source: by author

2 Transnational Cooperation

A Programme and a Project Perspective

As a considerable share of programmes and projects funded by the EU is based on transnational cooperation as the dominant principle of action, these programmes lend themselves to the study of transnational learning and knowledge development. Among these, the transnational strand of the INTERREG programmes, which fund actions for regional development, probably has one of the most ambitious transnational approaches. Its regulatory frame is determined by EU Cohesion Policy, which with its main objectives and principles impacts on the characteristics, the setup and the operation of individual projects. At the project level, the transnational dimension manifests itself in the type of partnership and the chosen topics and objectives, but also in terms of their management and operation as well as any investments made. The complexity and diversity of transnational cooperation together with the requirement to adapt to their specific framework (in this case the INTERREG programmes) impact on – and sometimes considerably challenge – their ability to produce results.

2.1 The Programme Perspective: Transnational Cooperation under the European Territorial Cooperation Objective

As a specific EU policy instrument, transnational cooperation has been around for some 30 years, but it has also occurred in many other contexts where people work together on joint issues and exchange knowledge across national borders, be it in publicly funded initiatives or in private companies joining forces. These different contexts of transnational cooperation influence its character and objectives while many of its expected benefits and related challenges are highly comparable.

Many EU programmes are based on the idea of European stakeholders working together in a certain field. Although most of these programmes are built on the partnership principle, working across national borders is often a possible option rather than a necessity. EU programmes promoting transnational cooperation include those in the field of research (HORIZON 2020) as well as the former 'Community Initiatives' in the area of regional (INTERREG, ESPON) and urban development (URBACT), which today continue under the 'European Territorial Cooperation Objective'. Of these programmes, *the transnational INTERREG programmes are a particularly interesting case for the study of transnational learning and knowledge development due to their complexity in content and methods and diversity of stakeholders involved.* This complexity and diversity evoke some of the main challenges that transnational learning processes face and have to deal with.

Transnational cooperation is an integral part in INTERREG, ESPON and URBACT, while HORIZON 2020 also allows non-partnership projects (that is, with only one beneficiary) or national cooperation. While ESPON funds transnational research projects in a top-down approach where partnerships apply under set terms of reference, URBACT and INTERREG follow a bottom-up approach. Projects are developed by applicants and then put forward for funding to the programme bodies. Both programmes promote an action- and implementation-oriented approach with a mix of project partners from research, administration and policy. The programme that most specifically focuses on transnational cooperation as its main operational principle is the INTERREG strand B. This strand comprises 15 different transnational cooperation areas, which all support the development of innovative measures in regional development. Although the other two programme strands (A and C) are based on cooperation across national borders (cross-border and interregional), cooperation in the B strand is particularly challenged by complexity. Cooperation in the cross-border strand is characterised by the removal of concrete barriers to transport, education, healthcare, employment and innovation in border regions, often in the shape of creating and improving infrastructure. Cooperation in the interregional strand, on the other hand, is characterised by a more open exchange of experience and knowledge sharing. The B strand is much less focused on infrastructure and regional benefits and more on the creation of outcomes and results of transnational benefit. The strand is characterised by often more vague or complex cooperation results such as strategies, concepts, plans and decision-making. Strand B, it can, therefore, be argued, is more process-oriented than the A-strand. Moreover, cooperation in strand B is methodologically challenging due to a requirement for transnational working approaches that include joint design, development and implementation and the consideration of the territorial dimension. At the same time, it requires more tangible outcomes and results than those typically produced under the C strand.

The institutionalisation of cooperation in the form of programmes makes EU Cohesion Policy particularly suited for the analysis of learning processes in transnational cooperation. The transnational strand of the territorial cooperation objective (strand B) is a particular interesting study area of transnational cooperation with its dual challenge of providing 'concrete' results, i.e. beyond networking, while in its subject matter not solely addressing tangible benefits at regional level, such as local infrastructure.

There are a few *key principles and objectives* that guide the content, management and operating mode of the EU Cohesion Policy programmes in general and of the programmes of the European Territorial Cooperation Objective in specific and thereby also more or less directly impact on individual cooperation projects. Understanding these helps to decode some of the basic characteristics of INTERREG programmes and projects, which, again, have considerable bearing on learning and knowledge development in projects.

Objectives and Key Principles of EU Cohesion Policy

The *main objective of EU Cohesion Policy* is the reduction of regional disparities throughout the EU. This objective was first introduced in 1975 with the
establishment of the European Regional Development Fund (ERDF), although
some regions had received financial support through the European Social Fund
(ESF) before this point in time. Together, these two funds make up the 'Structural
Funds'. Besides promoting social, economic and territorial cohesion, the Structural
Funds have been aiming at realising the objectives of the Community, especially
those of the Lisbon Strategy for Growth and Jobs (European Council 2000), the
Gothenburg Strategy for Sustainable Development (European Council 2001) and
more recently, the Europe 2020 Strategy (CEC 2010).

The reform of the Structural Funds of 1988 can be understood as a reaction
to the experience that market integration through the Single Market Programme
would not 'naturally' lead to economic convergence between European regions.
The reform introduced the four *key principles to allocation and management*
of relevant programmes that are still valid today: multi-annual programming,
concentration on the poorest and most backward regions, additionality and the
involvement of regional and local partners.

Due to a lack of formal EU competency in the field and as opposed to the
regulatory policy areas of the EU (such as in environment policy, food safety and
many others), the *governance mode of Cohesion Policy* is less characterised by
directives and more by funding schemes that bring added value to actions on the
ground and help to finance concrete projects through its 'unique model of multi-
level governance [that] involves local and regional actors in the policy design and
delivery, bringing in more efficiency and local knowledge' and that 'makes people work together through numerous cross-border and transnational programmes
and networks' (CEC 2008: 4). Thus, much of the EU's regional policy in general
and of its funding programmes in particular includes strong bottom-up aspects.
The ERDF Regulation and the Common Provisions Regulation define a general
frame for programme funding (Regulation (EU) 1301/2013, Regulation (EU)
No 1303/2013). This is then given life by individual 'Operational Programmes'
(OP), which determine thematic priorities and guide the selection of projects.
Policy implementation is thereby based on interactive relationships between
various levels.

Objectives and Governance Mode of the INTERREG Programmes

The programmes of the European Territorial Cooperation Objective that were first
introduced as a 'Community Initiative' in 1990 by the name of 'INTERREG',
focus on common challenges and opportunities and experiment with new
approaches and concrete actions across national borders. The *general objective
of the INTERREG programmes* is to provide a framework for joint action that
advances Cohesion Policy's overall objectives of the harmonious economic,
social and territorial development of the Union. During the first funding period

(1990–1993), 31 Operational Programmes promoted cooperation across internal borders with a thematic focus on tourism, rural development, transport and communication, subject matter expert (SME) support, local infrastructure and pollution prevention. The following funding period (1994–1999) introduced the three different 'strands': strand A continued the cross-border cooperation, strand B supported the completion of energy networks in southern Europe and strand C – which was added in 1997 – introduced the concept of transnational cooperation in specific cooperation areas. INTERREG III followed its predecessors and switched transnational cooperation to strand B while the new strand C was dedicated to Europe-wide cooperation activities. This funding period paid stronger attention to comprehensive cross-border and transnational strategies, to the European added value and 'truly joint structures' for programme preparation, management and implementation as well as complementarity to other programmes and initiatives (CEC 2004). From then on, INTERREG has been based on the following *key principles* (CEC 2000):

1 *Joint programming*: regions draft programmes that describe their joint development strategy and that clearly show the cross-border/transnational character of investments and the added value from cooperation. All measures are then developed on the basis of these programmes.
2 *Partnership and bottom-up concept*: only measures that are developed in partnership are supported.
3 *Complementarity*: with the main interventions of the Structural Funds.

Since the 2007–2013 funding period, INTERREG is no longer a Community Initiative but became part of the mainstream objectives of the Structural Funds as the European Territorial Cooperation objective (ETC). For the new 2014–2020 funding period, ETC programmes are based on their own regulation (Regulation (EU) 1299/2913) and available funds aggregate to €10.1 billion, equivalent to 2.9% of the Structural Funds (website DG Regio[1]). The larger part of these funds (ca 74%) is employed for cross-border cooperation, while some 20% of the funds are available for transnational cooperation and some 6% for interregional cooperation (Regulation (EU) No 1299/2013).

The *governance mode of the INTERREG programmes* is based on 'joint management' where member states establish Joint Secretariats (JS) responsible for the programmes' daily management, progress monitoring, regular reporting to the Commission and assistance to the Managing Authorities. Moreover, ETC programmes have Steering and Monitoring Committees at their disposal, which consist of member states' representatives. Their responsibility includes the assessment of project applications, considering not only project quality but also their strategic fit to the programme.

---

1 http://ec.europa.eu/regional_policy/en/policy/cooperation/european-territorial/ (accessed on 23 June 2015).

INTERREG's Focus on Innovation through Transnational Cooperation

Similar to other EU programmes, transnational cooperation in the INTERREG context focuses on supporting innovation. This focus on innovation (see box below) spotlights the programmes' knowledge dimension, as innovation processes are inevitably linked to the development and advancement of knowledge in one or the other way.

The diversity of regional policies and their varying frameworks, diverse institutional constellations and the range of experience with different models and instruments, approaches and solutions provides a rich source of experience from which to draw in innovation processes. Transnational cooperation is expected to lead to highly dynamic settings for creative experiments, in which a multitude of ideas and approaches are tested between different contexts (McFarlane 2006). Such testing and experimenting is difficult for regions to achieve single-handedly, but in cooperation they can examine the suitability of individual concepts and approaches under diverging conditions. INTERREG programmes support the transfer of both existing and emerging innovative concepts and instruments in networked rather than linear and uni-directional transfer situations.

Innovation as a Policy Priority in Transnational Cooperation

When the INTERREG programmes were integrated into the mainstream Structural Funds policy after 2007, they also had to make sure to contribute to the relevant overall policy guidelines, such as the renewed Lisbon Strategy for Growth and Jobs. The Community Strategic Guidelines for Cohesion (European Council 2006a) outlined that the programmes supported by cohesion policy, such as the INTERREG programmes, should target their resources according to the new Lisbon strategy, which included a focus on innovation, entrepreneurship and the growth of the knowledge economy. Following this, the general provision on the European Regional Development Fund determined four thematic priorities for the transnational strand of the INTERREG programmes, one of which was a focus on innovation (European Council 2006b). Many transnational programmes then also defined innovation as cross-linking all other thematic priorities as a so-called horizontal principle underlying all projects. Similarly, the Regulation for the ERDF during the 2014–2020 period require the ETC programmes to contribute to the Europe 2020 strategy for smart, sustainable and inclusive growth (Regulation (EU) No 1303/2013). In the new INTERREG programmes, innovation is one of 11 possible thematic priorities and has been selected by most of the transnational programmes as a funding priority.

In theory, transnational INTERREG projects cover the whole spectrum of the innovation value chain from research and development over testing and demonstration to commercialisation and market introduction. However, classic research-intense product innovation is not their core area; rather, it is practical solutions for regional challenges. Although some product innovation is funded, transnational INTERREG programmes are geared towards more integrated approaches to regional capacity building, process innovation (particularly in the public domain) and the establishment and advancement of transnational innovation clusters. In practice, relevant learning outcomes as the result of cooperation processes include changed routines, management structures and policy styles, new conceptual understandings, adapted procedures, methods and policies and new regional partnerships (Lähteenmäki-Smith and Dubois 2006; Böhme et al. 2004). Learning about innovative instruments and methods implies that stakeholders' own measures and instruments are improved and adjusted as a result of relevant knowledge transfer (so-called 'single-loop learning'; see section 3.1.2). Additionally, more innovative outcomes can be achieved when strategic and structural changes are carried out and existing systems are questioned (so-called 'double-loop or systemic learning').

Objectives, Key Principles and Governance Mode of the Transnational Strand of INTERREG

Strand B of the INTERREG programmes is dedicated to the promotion of transnational cooperation. Originally, its *objective* was 'to promote a harmonious and balanced development of the territory of the European Union; to foster transnational co-operation within a common framework in the field of spatial planning by the member states, regions and other authorities and actors; to contribute to improve the impact of Community policies on spatial development and to help member states and their regions co-operate on a pro-active approach to common problems, including those linked to water resource management caused by floods and drought' (website DG Regio[2]). The spatial planning dimension of the programmes may have been surprising considering the lack of EU competence in the field. Still, they were explicitly intended to be one of the key operational mechanisms for implementing the intergovernmental initiative 'European Spatial Development Perspective' (EDSP) (CSD 1999). Since the integration into the mainstream objectives of the Structural Funds after 2007, strand B has moved away from thematic priorities influenced by ESDP concepts in favour of the Lisbon and Gothenburg agendas and later also of the Europe 2020 strategy. Thematic priorities are now chosen from a list of 13 possible thematic objectives applicable to all Cohesion Policy funds, including research and innovation, low-carbon economy, climate change, environmental protection and sustainable transport (Regulation (EU) No 1303/2013). Still, transnational cooperation projects are supposed to contribute to an integrated territorial development, which also includes the consideration of their potential impact on different sectors in their respective territories.

Following the transnational strand of INTERREG through its funding periods since 1995 demonstrates the evolution of its requirements/characteristics (see Figure 2.1). While INTERREG IIC is widely seen as having been of experimental character and lacking a strategic dimension and concrete results (Pedrazzini 2005;

---

2 http://ec.europa.eu/regional_policy/archive/interreg3/inte2/inte2c.htm (accessed on 5 March 2014).

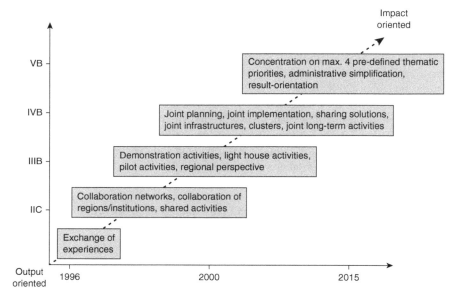

Figure 2.1 Transition of activities in the transnational INTERREG programmes

Source: adapted from Byrith 2009

LRDP 2003), quality increased towards the IIIB period. Projects were supposed to deliver more tangible results and durable structures (CEC 2000) and in many cases, this has also been achieved (Panteia et al. 2010c). The increased relevance of 'pilot projects' can directly be linked to the call for more tangible results and increased the programmes' visibility on the ground. On the other hand, this development led to project results mainly visible at the local level and to doubts about the projects' transnational character. This again fuelled calls for projects with a stronger strategic and transnational dimension for the IVB period. Programmes tried to pick this up by introducing 'strategic initiatives', 'strategic projects' and 'clusters' as well as by a much more sophisticated elaboration of the concept of 'transnationality', such as in the concept of 'transnational investments' (NWE 2010) during the INTERREG IVB period. The INTERREG VB programmes (2014–2020) now focus on a stronger result-orientation in general and provide projects with a set of tools that may support them in achieving their objectives. These include the requirement for projects to focus on achieving some kind of 'change', to contribute to certain set 'programme results' (Interreg CENTRAL EUROPE 2015), or to work with dedicated risk assessment (NWE programme).

This evolution of the transnational INTERREG programmes took place parallel to a budgetary and regulatory development that increased programme budgets while simultaneously raising and substantiating requirements and expectations. In essence, this resulted in a perpetual alignment with the EU policy framework.

Still, programmes only provide the framework for practical project work and projects are the final means to fill the above-mentioned requirements with life. Projects have therefore over time also increasingly been challenged by the expectation to effectively initiate processes and produce results of truly transnational character, which again require target-oriented knowledge development and learning. It is therefore time to switch attention away from the overall programmes to transnational projects themselves.

2.2 The Project Perspective: Characteristics and Challenges of Transnational INTERREG Projects

Of all the characteristics of transnational INTERREG projects, *transnational partnership* is probably the most important. A transnational partnership usually involves partners from at least three different countries of the cooperation area. The Northwest Europe programme also allows partnerships based in only two countries; in practice, however, projects in this cooperation area typically include partners from more than two countries. As the case studies presented later are all based in the Northwest Europe programme (NWE), the following characterisation of transnational INTERREG projects can slightly deviate from those of other programmes.

The NWE cooperation area, which consists of the UK and Ireland, Luxembourg, Belgium, Switzerland and large parts of the Netherlands, together with the northern part of France, five German federal states as well as parts of Bavaria, slightly differs from other cooperation areas by traditionally disposing over a higher overall programme budget. In addition, project budgets are higher than in other cooperation areas (€3.28m on average during the IVB period[3]) and project lifetimes longer (54 months on average during the IVB period[4]) (BBSR 2009). Higher project budgets are also reflected in comparably high project investments. During the INTERREG IVB period, participation was highest from the UK and Belgium, but Dutch, German and French partners were also highly engaged (BBSR database[5]).

The Institutional Structure of Transnational Projects

The *institutional structure* of projects is typically characterised by a high participation from regional and local authorities, but NGOs are also active participants. Other partners include universities and research institutes, national and sub-national authorities, private organisations and foundations. Moreover, projects need to be of cross-sectorial and inter-disciplinary character, which means that even if partners have similar institutional backgrounds, their professional backgrounds can

---

3 For comparison: the average budget for projects in the five cooperation areas that Germany participates in is 1.61m EUR.
4 For comparison: the average lifetime of projects in the five cooperation areas that Germany participates in is 37 months.
5 https://www.bbr-server.de/interreg/ (accessed on 20 May 2015).

differ considerably. Finally, project complexity is even more increased by the demand for 'vertical integration' or the inclusion of administrative levels capable of implementing project results in relevant policies. As competencies of public bodies vary throughout Europe, including the 'relevant' institutions often implies a certain diversity of the spatial and administrative levels involved. Projects are thus characterised by high institutional, cultural and disciplinary diversity and are often of multi-level character.

Every project is based on the *partnership principle* and includes a 'Lead Partner', who is ultimately responsible for managing the project on behalf of the partners. Each partner has a specific role to play in the project as identified in the project application. Projects are 'expected to work transnationally throughout each phase of the project (joint design and development, decision-making, implementation, evaluation and dissemination)' (NWE 2015: 13).

Partnerships apply for funding in a bidding process in one of the 'application rounds'. The Joint Secretariat (JS) assesses applications according to an assessment scheme based on eligibility criteria and – if these are passed – selection criteria. Based on this pre-assessment, the Programme Steering Committee (PSC) decides which projects ultimately receive funding. In most other cooperation areas, the programmes' Monitoring Committees (MC) are responsible for project approval. During the INTERREG IVB period, the PSC in Northwest Europe (1) rejected a project, (2) approved a project, (3) conditionally approved a project, or (IV) referred a project back to the next round. Unconditional approval was the least common decision.

Cooperation Topics of Transnational Projects

Possible *cooperation topics* are defined by the strategy and priorities of the relevant Operational Programme. As mentioned above, there has been a shift away from spatial planning concepts in order to answer to the Lisbon, Gothenburg and Europe 2020 strategies. However, the programme still requires projects to contribute to the development of the wider cooperation area and many funded projects still include aspects of urban and regional development. In terms of projects' demand, during the NWE IVB period the priority focusing on environmental aspects was highly popular and projects dealing with aspects of climate change were prominent. In addition, both the innovation and the urban and regional development priorities were sought after, whereas similar to other cooperation areas, the transport priority had more difficulty attracting projects. However, the urban and regional development priority suffered low approval rates due to a lack of transnationality in project approaches (BBSR 2009).

In essence, the transnational project character refers to problems that cannot be solved efficiently by individual states or regions and solutions that can best be jointly developed by organisations in different member states (NWE 2010: 25). In an approach to conceptualise the transnational character of its projects, the NWE programme introduced a distinction between 'common' and 'transnational' issues during the INTERREG IIIB period (NWE 2003). Although this definition is no longer officially used, it still provides a valuable basis for understanding two quite distinct project types. Projects based on a 'common issue' justify cooperation by an issue 'faced by several cities and regions in various locations across the European territory, which could be or has been tackled at the local, regional or national level, but for which transnational cooperation would bring more innovative and efficient solutions' (ibid.: 22). Theoretically, these can also be solved without transnational cooperation. Projects based on a 'transnational issue' deal with issues that affect 'a transnational area across national and regional borders, which cannot be tackled adequately at the local, regional or national level and which requires transnational cooperation' (ibid.).

Project Strategies of Transnational Projects

The projects' transnational character applies not only to their partnership and the topic they choose, but also to the way they operate and are managed, as well as to any investments made. Both cooperation topic and actions need to be of relevance for the whole programme area. In practice, this often seems to be challenging for projects, and during the IIIB period, many projects did not have much more in common than a joint title and involved rather unconnected measures and investments (BBSR 2009: 32). The IVB period put more emphasis on projects actually working out their transnational approach.

The basic *project strategy* of INTERREG B projects is guided along 'work packages', in which partners group up into teams. A large majority of projects builds on the concept of 'pilot projects' or 'transnational investments'. The term 'pilot project' refers to an activity planned as a small-scale test or trial that is meant to demonstrate the usability of new concepts for a later large-scale implementation and is widely used in EU-programme speech.

Transnational Project Results

In terms of *project results*, transnational INTERREG programmes draw a distinction between the dimensions 'output', 'result' and 'impact'. 'Output' relates to project activities and is usually measured in physical units (for example number of schemes implemented). These can take many shapes and can range from transnational studies over development concepts and action plans to concrete investments. Transnational investments need to show a clear spatial dimension, produce functional or physical relations, be of exemplary character and offer transferable actions and methods (BBSR 2009; NWE 2010). During the evolution of the transnational programmes, attention towards the implementation of the projects has been increasing, and studies, for example, are now only funded if directly followed up by concrete measures or investments.

'Results' relate to the direct and immediate effects of these outcomes. They provide information on changes, for example on the capacity or behaviour of beneficiaries (for example the number of institutions with improved capacity). Results can include the introduction of harmonised standards and transnational brands or marketing strategies, the implementation of EU guidelines and their integration in the local and regional context, influence on political agendas as well as the establishment of regional networks and institutions (BBSR 2009).

'Impact' then leads back to the overall project objectives and provides information on long-term effects, which are not necessarily of physical nature (NWE 2010). In the context of transnational cooperation, project impacts are often rather indirect and long-term, but they can also go well beyond the original project objectives (Hübner and Stellfeld-Koch 2009).

A variety of studies and evaluations have analysed the direct results, effects and added value of transnational cooperation in the transnational INTERREG context since the early 2000s and have identified a number of challenges specific to this context. Many of these had a focus on spatial planning issues.[6] In general, researchers argue that causal relationships between learning, policy outcomes and change are not easy to make. If at all, these studies often only found moderate effects of INTERREG B programmes and projects. They point to the fact that many projects do not fully use the potential of their transnationality, interdisciplinarity and practical orientation. In some cases, this has even led to questioning the legitimacy of whole programmes. Overwhelmingly, research has shown that the outcomes of INTERREG can less often be found in tangible results, but more often connected to 'learning effects'. However, the use of the term 'learning' varies between studies and is dominated by an understanding of learning as the dependent variable of cooperation processes (as a result of cooperation rather than an impact on the cooperation result). Occasionally the use of the term strongly diverges from an understanding of learning as a ubiquitous and integral part of human activity that cannot be avoided as such (Elkjaer 2008) and as the basis for the more tangible results and effects.

As reasons for unsatisfactory results of INTERREG B projects, researchers identify the high administrative effort for the projects as well as the fact that many participants are not primarily interested in cooperation and only make use of INTERREG as a funding source for their local projects. Moreover, the focus of many projects on 'common' rather than 'transnational issues' and low cooperation intensity prevents them from developing a direct transnational impact. A variety of factors potentially influencing transnational cooperation and learning have been identified. These, however, have their shortcomings as they are based on (1) a strong focus on outcomes (rather than project processes), (2) a focus on planning issues, which are not topical to all INTERREG B projects and (3) a lack of conceptualisation of the assessment of ('soft') learning outcomes and effects. Although newer studies have shifted the focus from physical outcomes to the role of 'softer' and

---

6 For example Dühr et al. 2010; Panteia et al. 2010; Wink 2010; Hübner and Stellfeld-Koch 2009; Mirwaldt et al. 2008; Stead et al. 2008; Bachtler and Polverari 2007; Colomb 2007; Dühr and Nadin 2007; Peterlin and Kreitmayer McKenzie 2007; Waterhout and Stead 2007; Dabinett 2006; Lähteenmäki-Smith and Dubois 2006; Mairate 2006; Dabinett and Richardson 2005; Giannakourou 2005; Janin Rivolin and Faludi 2005; Peddrazini 2005; Zaucha and Szydarowski 2005; Böhme et al. 2003; Hassink and Lagendijk 2001.

more long-term effects, the evaluation methods used for assessing INTERREG B projects' results and impacts do not always seem to have been adapted to this fact.

Challenges of Transnational Cooperation Projects

This book introduces a perspective on transnational cooperation that considers project strategies and processes, which allows the identification of additional factors that impact on transnational learning processes. Transnational projects are based on the general idea that the diversity of local and regional development processes acts as a knowledge source for the design of innovative policies and strategies and that lessons can be learned from the experience of actors from other national contexts. However, the preconditions for knowledge sharing, transfer and development in the highly complex partnership structures of transnational cooperation projects are challenging. In order to deal with these challenges successfully, considerable effort and notable resources are needed. Methods for transnational learning can support these processes. The book argues that projects' softer and more long-term effects are difficult to identify and understand without projects and programmes adopting a systematic approach to learning processes.

Examples of challenges and obstacles to project learning processes that ultimately support the development of satisfactory project results include:

- Much of the knowledge that is potentially valuable to share with transnational partners does not exist in the form of explicit, codified knowledge but rather as tacit and context-dependent knowledge that is hard to formulate.
- The transferability of experiences from one context to another is limited, particularly for actors based in different 'models of society'.
- Related to the previous point, a simple copying of experiences and 'best practice' (or 'good practice') is usually not realistic. Instead, these require further processing and adaptation to fit different preconditions and resources.
- The concept of 'best/good practice' is widely used in transnational cooperation. 'Best/good practice' as an input to knowledge production is related to the above-mentioned challenge of context dependency. But 'best/good practices' are also very often the output of transnational cooperation projects. This leads to the aspect of quality assessment, as the distinction of 'best' or 'good' implies some kind of preceding benchmarking and or evaluation process. Due to the strong context dependency of these practices, however, processes of comparison can prove highly challenging and require procedures that establish a comparable basis.
- As individuals are the agents of learning, learning first and foremost takes place at the level of individual partners. These can draw their individual lessons from the transnational exchange of experiences and practices. However, in more action- and implementation-oriented programmes such as the transnational INTERREG strand, projects are expected to develop new concepts, strategies and solutions to pressing problems (development

of *new* knowledge). This calls for a group effort and goes beyond individual learning. Group or project learning demands structures for the exchange, processing, monitoring, assessment and evaluation of knowledge and experiences that originate from a variety of sources. Again, considerable effort and resources are needed to allow the development of joint knowledge based on individual experiences.

- In programmes such as the transnational INTERREG strand, where the aim of transnational exchange is to solve common, practical local and regional challenges, project partners are expected to implement newly gained knowledge in decision-making structures. This puts extra emphasis on the organisational context of the project itself and requires new knowledge to leave the isolation of the project and enter the world of practice of the recipient organisations.
- In projects such as INTERREG projects, the diversity of project partners is remarkably high. In addition to partners' transnational background, they often also differ in terms of being interdisciplinary (research, administration, NGO), multi-level and different geographical levels they represent (local, regional, national, and international).
- Another speciality of INTERREG, but also other transnational projects, is the concept of '*pilot projects*'. If pilot projects are used as individual tests for a given problem in a transnational lab and evaluated so that the project can settle on the best solution(s), they need to have a functional relationship to each other and to the overall project objective.
- One particular challenge of INTERREG B projects, due to their practical orientation, is the *development of joint lessons and knowledge* from a limited selection of individual implementation projects and thus the ability to jointly make sense of case-based knowledge.

3   Three Keystone Concepts
Learning, Knowledge and Cooperation

The following chapter discusses the three main aspects that are of relevance when analysing transnational learning and knowledge development, namely conceptions of 'learning', 'knowledge' and 'cooperation'. The section on 'learning' takes a look at how both psychological and organisational studies approach the subject. Particular attention is paid to learning theories that include the notion of 'social learning', as learning in transnational projects is learning in interaction. Similarly, the concept of 'knowledge' is discussed in relation to different ways of conceptualising and describing knowledge. The concepts of learning and knowledge are closely interlinked in a mutually reinforcing process: 'while learning (the process) produces new knowledge (the content), knowledge impacts future learning' (Vera and Crossan 2008: 131). Knowledge is thereby both a resource for a process that forms the basis for expertise, skills and competence (input) and a product that leads to new developments (output), but its exchange and development are also relevant to the process of learning (Humpl 2004). A short section on cooperation follows that provides an overview of the main characteristics and challenges of cooperation in international, inter-organisational and inter-disciplinary projects.

3.1 Learning

Throughout the history of research into learning, various theory strands have attempted to interpret learning, but no single theory has been able to account for the multiplicity of processes taking place in human minds. Thus, 'each theory describes the key features of learning as the theorist defines them and focuses on identifying factors that will lead to those outcomes' and 'each theory begins with an assumption or a set of assumptions about the nature of learning and then proceeds to develop a set of principles consistent with the assumptions' (Gredler 1997: 15ff.). Since the so-called behaviouristic approaches of the late nineteenth and the early twentieth centuries, learning theories have developed in various directions with numerous paradigms in relation to learning processes, but behaviouristic approaches still form one of the main strands.

The object of learning can focus on content and thereby relate to instrumental learning. It can also focus on relations, so-called 'communicative learning', when learning leads to a better understanding of the interests and values of others, to the building of relations, enhanced trust and cooperation (Diduck et al. 2012).

Two disciplines have extensively dealt with learning concepts and processes: psychological studies and their understandings of learning in behaviouristic, cognitive and constructivist ways and continuing with organisational studies and their conceptions of organisational and inter-organisational learning that provide categorisations of learning units and levels.

3.1.1 Learning Explained in Psychological Science

Today, the major strands of learning theory include behaviouristic, cognitive, social-cognitive and constructivist theories. *Behaviouristic theories* stress learning from interacting with the environment, but see learning processes as rather mechanistic and more influenced by the environment than by learners themselves. The individual and their cognitive processes are treated as 'black boxes', leaving learning processes largely unexplained.

Cognitive theories integrate learners' perceptions, analyses, plans and choices and move the 'black box' of the mind with its processes of thinking, knowing and memorising into focus. These theories do not interpret changes in human behaviour merely as learning outputs, but as a clue to what is going on in the individual's mind. Effective learning in this understanding depends on realistic perceptions of the challenges and how fast learners can discover changes. As part of cognitive learning theories, *social-cognitive theories* emphasise learning from other members of a social community and are of particular interest to cooperation projects. Social-cognitive theories not only look at individual learning processes in terms of external stimuli and internal forces but also include the interaction between environment and humans. Learning is understood as an active and controlled process of assimilating experiences (Verres 1979). The relevant cognitive procedures are of particular interest, with a distinctive focus on the human ability for vicarious learning by witnessing others' experiences.

Constructivist theories emphasise the 'construction' of knowledge and its subjective character. Constructivism recognises that 'knowledge is not part of an objective, external reality that is separate from the individual' but that it is a 'human construction' (Gredler 1997). Learning, in this perspective, is a contextualised process in which knowledge is not so much obtained as constructed, based on a person's prior knowledge and experiences and his or her propositions of the environment (Merriam et al. 2007). Coinciding with a social constructivist turn in social sciences, social learning approaches developed as part of this theory strand. By acknowledging that people interact with others, themselves, contexts and artefacts, the focus of learning is moved away from the individual mind to the social sphere of interaction, activity and practice (Cook and Brown 1999). As each person has different interpretations and constructions of knowledge processes that guide their action, social learning approaches emphasise the need to test one's theories and understandings through social interaction and to inquire into any relevant differences of perceptions among the different actors involved

(Friedman 2001). In this context, Vygotsky's social development theory (1978) with its strong focus on learners' cultural-historical contexts stresses the relevance of an individual's cultural context on their perception of reality, a notion of relevance in transnational settings.

3.1.2 Learning Explained in Organisational Studies

Organisational studies have a long history in conceptualising learning processes and effects taking place alongside practical organisational activity. Classic organisational learning theory focuses on intra-organisational aspects, which is relevant when attempting to understand how the lessons learned in transnational projects are later applied and implemented in the host organisations. Although it is mainly concerned with learning *within* organisations, some of its basic concepts are also of relevance for learning *between* organisations.

Organisational learning takes place 'when members of the organization act as learning agents of the organization, responding to changes in the internal and external environments of the organization by detecting and correcting errors in organizational theory-in-use, and embedding the results of their enquiry in private images and shared maps of organization' (Argyris and Schön 1978: 16). It is the impact on both thought and action stored in non-human repositories such as routines, systems, structures, culture, and strategy (Vera and Crossan 2008).

Organisational learning provides useful conceptualisations such as units, levels and depths of learning, as well as with concepts and models of learning processes. The latter allow going beyond traditional input-output considerations and shed light on learning triggers, supporters and phases.

However, applying the concept of organisational learning to settings of project learning between several organisations in a transnational context faces a variety of shortcomings. First of all, a general deficiency of organisational learning theory is a lack of conceptualisation, a widely accepted framework or even consistent terminology (Vera and Crossan 2008). Not all approaches are concerned with describing and analysing learning for the purpose of theory building; some are clearly of normative character and expect major transformations within organisations (Berthoin Antal 1998). The empirical basis of organisational learning is strongly characterised by research from Anglo-Saxon countries with a strong focus on the management level in firms. Moreover, many authors use the term 'organisation' as a synonym for 'company', whereas theorisations of learning in other organisational forms are still rare. In most of the literature, organisational learning is concerned with organisational reforms, privatisation or restructuring processes, often caused by external pressures like profit losses or economic crises. However, organisational learning is not limited to crisis management but refers to all aspects of an organisation: modifying product design, changing the organisational form or optimising processes (Humpl 2004).

In EU cooperation projects, learning processes are of a more thematic nature; often, people come together in order to find out more about a certain topic or area and how they can introduce or improve new features in their community or region.

Actors usually see a meaning in cooperating as such, not necessarily because they face strong pressure to act. Quite the opposite, as the economic crisis of the last years and political restructuring processes have shown, actors have been withdrawing from cooperation projects (particularly in Eastern Europe and the UK). This shows a different understanding of potential organisational learning, less as an option for learning and skills to master problems and more as learning on very specific topics that need to stand back in times of crisis.

Questions remain especially concerning the transferability of the concepts of organisational learning to project organisations due to the rather short-term objectives of projects (cf. Humpl 2004; Schindler 2000). While organisations are concerned with their long-term survival, projects are more oriented towards fulfilling their specifications as well as the needs of the relevant participants and stakeholders. This means that the limitation in time that projects face opposes basic ideas of organisational learning, which is aimed at securing long-term survival. Often the process of securing knowledge gains does not have priority in projects (see section 5.3.5). Thus, the presented models need to be treated with care when applied to short-term project settings.

Another reason for a critique of the concept of organisational learning is that much of the literature on organisational learning is based on concepts of individual learning. It is assumed that individuals learn – in a way, on behalf of their organisation – and the outcomes of these learning processes are then crystallised into organisational routines and memories and become organisational learning. Learning is regarded as a specific activity that needs to be initiated and stimulated and mostly takes place in case of problems occurring that need to be solved. Accordingly, it is assessed on the basis of the resulting change in organisational routines (Elkjear 2008). In sum, organisational learning understood in the context of individual learning theory is actually individual learning in organisations, which implies the challenges of transferring its outcomes into the organisational body.

Despite all the shortcomings of organisational learning literature, some of its basic concepts are of relevance to transnational projects. Beyond these, organisational learning theory has a lot to contribute to the understanding of the implementation of project results in the relevant home organisations, which – although not the main concern of this book – is highly relevant for ensuring their added value and long-term effects.

Organisational learning theory provides two categorisations of learning that help to conceptualise learning actors (units of learning) and different learning depths (levels of learning).

A Units of Learning

It is useful to differentiate between different units of learning. All learning is impossible without processes of individual learning and, up to the present day, many studies in organisational learning focus on processes of individual learning. At the same time, learning does not only contain individual, but also social elements. If learning takes place in a professional environment, where individuals are associated with organisations, learning also occurs at the level of the organisations involved. Although none of these perspectives can provide the full picture, the three levels can be seen as integrative to each other.

INDIVIDUAL LEARNING

Individual learning refers to personal learning processes, which – in the case of projects – take place among project participants representing different organisations. Organisational learning in the individual-centred view is considered to be the sum of the learning of individual members of the organisation (Knight 2002). When individuals leave, their knowledge will be lost for the organisation.

GROUP OR PROJECT LEARNING

Group learning takes place if learning happens in a social context (such as in a cooperation project) and a collective memory is assembled. Group learning is distinct from individual learning and often harder and more complex as it refers to interrelated activities among several people (Weick and Roberts 1993). Although the individual develops new knowledge, cross-fertilisation processes with other actors can occur and the further development, dissemination and use of the knowledge are linked to cooperation between actors. As decision-making processes are often collective procedures, the group or project as a whole is responsible for the outcome. Group learning is a process of cumulative knowledge creation, in which knowledge needs to be freely accessible and transferable among the actors involved (Capello 1999). These processes entail the sharing of experience and ideas between individual partners, the collective incorporation of the knowledge gained by individuals into collective capacity building processes and, based on this common starting point, collective decisions and initiative (Kissling-Näf and Knoepfel 1998). To avoid being lost, collective knowledge needs to be transferred to both individual and organisational knowledge. This level is the most relevant level of analysis for learning in transnational projects.

ORGANISATIONAL LEARNING

Although individuals are the agents of learning processes in transnational cooperation, they do not act as individual persons but are representatives of their organisation. After all, learning processes only integrate single, interchangeable representatives of each organisation, who take the knowledge with them when leaving the organisation. The learning effects of the participating institutions are critical for the following application and implementation of results (Sydow et al. 2004). Knowledge needs to be shared and disseminated to be incorporated in an organisation.

Organisational learning is considered to be more than the sum of the learning of individuals or groups that make up an organisation. Outcomes of organisational learning are modifications of an organisation's system, structures, procedures and culture that reflect and are reflected in changing patterns of action such as routines and strategies (Knight 2002). These learning processes are thus processes of transfer and institutionalisation. As new knowledge needs to be integrated into existing stocks of organisational knowledge, the transfer of knowledge into organisations is a process that requires creativity (Dierkes and Albach 1998).

Learning at the organisational level is in many cases more complicated and harder to achieve than at the individual and group level. In their study on learning in joint ventures, Inkpen and Crossan (1995: 595) find that 'while individual managers . . . were often enthusiastic and positive about their learning experiences, integration of the learning experience at the parent firm level was problematic, limiting the institutionalized learning'. Other authors come to similar conclusions, such as Hartley and Allison in their study on inter-organisational learning among local authorities (2002), or Easterby-Smith (1997) and Finger and Brand (1999). Although organisational learning is highly relevant for transnational cooperation projects, with its focus on implementing learning results from the individual and group level, it goes beyond the scope of this book.

B Levels of Learning

In their *Organizational Learning: a theory of action perspective*, Argyris and Schön (1978) propose a simple model to differentiate between two levels of learning: single-loop and double-loop learning. These are based on Piaget's distinction between accommodation and assimilation in learning: in assimilation processes, learners adapt to the current situation, while in accommodation processes they reinterpret the current situation and change their cognitive models accordingly (Piaget 1970).

SINGLE-LOOP LEARNING

Single-loop learning, also called 'adaptive learning' (Senge 1990), 'adjustment learning' (Hedberg 1981) or 'programme learning' (Dierkes and Marz 1998), describes a reaction to a changing external or internal environment that takes place within the existing framework of an organisation. During this reaction, existing norms and objectives are not questioned and the process is limited to detecting errors and correcting the applied 'theories-of-action'. In the context of organisations, single-loop learning is aimed at improving or modifying strategies and behaviour within an existing structure in a way that will best enable the organisation to achieve its goals (Dierkes et al. 2001). The experience and results of the learning process remain without influence on the behaviour and the underlying norms of the organisation. In the next context, the organisation will react and learn in the same way. Models of behavioural learning are clearly visible in this concept. Applied to units of learning discussed above, single individual learning refers to the learning of new facts and the correction of practice, while single collective learning corresponds to incremental changes in structures and collective rules (Newig et al. 2010).

Institutions are by most scholars considered to be 'path-dependent', which means that they confine choices to a limited range of possible alternatives and reduce the likeliness of path changes (Kazepov 2004). Routines and practices often work as stabilisation factors for existing institutional settings through positive feedback that encourages actors to continue on a given path and to focus on single alternatives (Pierson 2000). At the very basis of this lies the 'law of parsimony', which says that the premises of habits, routines and rules are not re-examined every time they are used (Bateson 1972). The path-dependency of institutions also implies that similar policies embedded in different institutional contexts can produce different impacts (Kazepov 2004).

DOUBLE-LOOP LEARNING

Double-loop learning, also called 'generative learning' (Senge 1990), 'turnover learning' (Hedberg 1981), 'experience learning' (Dierkes and Marz 1998) or 'environmental adaption' (Pawlowsky 1994), systematically draws on past experiences in which every context that is found is treated as a variant to a former context. An organisation 'recognises environmental changes that cannot be responded to adequately within the existing context of the organization or its proven strategies and behaviours' (Dierkes et al. 2001: 283). Not only do the 'theories-of-action' change, but also norms and objectives are adapted when changes prove incompatible with them. If behaviour is affirmed (for example by success), its repertoire will be stabilised, but if corrections become inevitable, the behaviour repertoire needs to be broadened or modified and existing priorities re-evaluated and changed. The dependence on certain repertoires is loosened while the dependence on experiences is strengthened (Dierkes and Marz 1998).

In addition, some authors extend this concept to the notion of 'learning-to-learn' and learning about the context of different situations, the so-called 'deutero-learning', 'turnaround-learning' (Hedberg 1981), 'problem-solving learning' (Pawlowsky 1994) or 'meta learning' (Dierkes and Marz 1998). In deutero-learning processes, learners expand their usual horizon, understand complex contexts and question implicit background assumptions on which their knowledge is based (Dierkes and Marz 1998). This requires people to become aware, understand and question the processes of adaptation and change as described above.

Applied to the units of learning, double-loop learning at the individual level refers to changes in assumptions and routines, while double-loop learning at the collective level relates to fundamental changes in structures and collective rules (Newig et al. 2010).

Obviously, which learning is appropriate in a given situation depends on the context. Minor environmental changes can be countered with the first two forms of learning while fast and dramatic changes may require learning of the third type. In real life, it is not always easy to differentiate between the three types, as they are highly dependent on a subjective interpretation of a particular moment or event. Moreover, the classification focuses too narrowly on problem detection and correction as a primary way of learning and is thus reactive rather than creative.

On the contrary, learning can also be triggered by opportunities, creativity or the acquisition of new knowledge (Dierkes et al. 2001).

Child and Faulkner (1998) further developed this rather theoretical approach into one applicable to real-life situations. In their view, the object of single-loop learning broadly corresponds to technical knowledge as it is mainly based on codified knowledge, whereas double-loop learning corresponds to the acquisition of systemic knowledge (changes in the organisational system). Deutero-learning, finally, refers to changes in managerial mindsets. The authors underline that in cooperative settings, there is a potential to learn at all of the three levels at the same time. Through routine or 'technical learning', cooperation can provide direct access to improved techniques and specific technologies and through 'systemic learning' it can facilitate the transfer and internalisation of new systems. Moreover, through 'strategic learning' cooperation can improve partners' capabilities as the cooperation opens a door to new strategic possibilities.

3.2 Knowledge

The term 'knowledge' has been discussed with varying philosophical and conceptual viewpoints in different theoretical debates so that no unified understanding and definition exists. While positivists argue that knowledge is the comprehension of an objective reality, post-modernists dispute this and stress that all meanings are context-specific. The former notion has been increasingly criticised and in this respect, the work of Polanyi (1967) has been influential, which shifted the focus to knowledge as an activity to the process of knowing. In the organisational field, Nonaka and Takeuchi (1995) developed particular aspects of organisational knowledge and set the standards for the newly emerging field.

From a project perspective, three dimensions of knowledge can be distinguished (Humpl 2004): the resource-oriented perspective understands knowledge as relevant for implementing an action, to reach the project's objectives and to form newly generated knowledge in terms of solving the problem. The result-oriented perspective sees knowledge as being included in the project result. Finally, the process-oriented perspective pays attention to the process of project development, which may impact on existing knowledge. Taken together, the three perspectives help form a definition of 'project knowledge' as a resource for targeted and rational action in a project, which can be found in the project's result and that is subject to changes during the process of the project's execution.

Another relevant term in this context is that of 'experience', which in daily use is often applied as a synonym for knowledge. When knowledge relates to content, having experience can be one way of acquiring knowledge and thus one specific way of learning (also called 'experiential learning') as opposed to academic learning. Experiences are linked to people, dependent on situations and contexts and based on perception (Humpl 2004). In the following, learning is understood as a ubiquitous and integral part of human activity that is based on making sense of experiences by creating new knowledge and taking over existing knowledge (Elkjaer 2008; Albach 1998).

Various attempts have been made to further conceptualise and categorise knowledge. Maybe the most obvious categories of knowledge are related to its *content*, such as the distinctions by Lane et al. 2001 (managerial knowledge, technological and marketing expertise) or Child 2001 (technical, systemic[1] and strategic[2] knowledge).

Lubatkin et al. (2001) distinguish knowledge types according to the process of individuals acquiring knowledge. First, they argue, an individual becomes familiar with an information domain and its semantics and acquires so-called 'know-what' (or 'declarative knowledge'). The individual then starts to form cognitive links between cause and effect through study, training and experimentation and develops a more specialised form of knowledge, so-called 'know-how' (or 'procedural knowledge'). With more exposure to the information domain, the individual builds up a solid expertise or knowledge structure for a range of information domains, so-called 'know-about'. These domains are interlinked and cognitively structured by the individual's 'information domain hierarchy' in the so-called 'know-why'.

The differentiation made by Polanyi (1967) into tacit and explicit knowledge and conceptualisations of the influence certain knowledge characteristics have on knowledge transferability are of particular relevance for transnational cooperation projects as they touch upon major challenges in exchanging knowledge in cooperative settings.

A Tacit and Explicit Knowledge

The distinction between tacit and explicit knowledge is one of the most widely used in literature and can be traced back to Michael Polanyi (1967). In his often-cited book *The Tacit Dimension*, he describes how the fact that 'we can know more than we can tell' (ibid.: 4) can make a distinction between different forms of knowledge. This is 'tacit knowledge', a range of mental models such as subjective perspectives, conceptions and intuition that can be used to make sense of something. Although tacit knowledge holds unique advantages for a project or an organisation, it is difficult to express in words and thus to transfer to others. It is particularly immobile between cultural boundaries (Makino and Inkpen 2008). Researchers do not reach a consensus as to whether tacit knowledge is conscious or unconscious.

'Explicit knowledge' (or 'codified knowledge'), on the other hand, can be expressed in words, numbers or symbols and can be transferred to others without difficulty by means of data, pre-determined procedures or universal principles.

Acknowledging the uncodifiable aspects of knowledge creation is important since it indicates that these processes are qualitatively different from the simple transfer of codifiable knowledge as information. In this spirit, learning involves more than simple transactions of information.

---

1 On relationships and roles.
2 Including reflexive cognitive processes to generate new insights.

The concept of tacit knowledge has become very popular since the mid-1990s and particularly since Nonaka and Takeuchi published *The Knowledge-Creating Company* (1995), in which they describe the transformation of tacit knowledge into explicit knowledge and vice versa in businesses (see section 4.2.2). They further develop Polanyi's concept by subdividing tacit knowledge into two dimensions: technical elements and know-how describe skills, while cognitive elements refer to perceptions of reality and future visions (ibid). They include, for example, the idea of knowledge creation through a transformation from tacit to explicit knowledge and the relevance of individual cultures to the construction of knowledge. The authors adopt a concept of knowledge that largely follows the definition as 'justified true belief' (Nonaka 1994). This definition emphasises both personal 'beliefs' and the 'justification' of knowledge, which adds a dynamic aspect to the concept of knowledge.

It has to be taken into account that Polanyi's distinction is not based on empirical findings, but on philosophical analysis and can thus pose challenges when applied in an empirical study. Moreover, it can be questioned how far tacit knowledge is translatable into explicit knowledge as Nonaka and Takeuchi imagine (Tsoukas 2008).

B Knowledge Characterisations Influence Knowledge Transferability

Salk and Simonin (2008) identify different knowledge characteristics and link these to their influence on knowledge transferability and storage. Of the knowledge characteristics, the tacitness of knowledge (see above), its complexity and specificity are regarded as particularly significant for its transferability, although the impact may vary between knowledge types (Simonin 1999a, 1999b). *Complexity* refers to the number of interdependent factors linked to the knowledge (for example routines, resources), which can limit transferability considerably. *Specificity* refers to the relative lack of the transferability of assets intended for use in a specific transaction. Other knowledge characteristics linked to a potential impact on the pace, depth and meaningfulness of learning are (Salk and Simonin 2008):

- *Validity*: accurate and reliable knowledge;
- *Novelty*: new or absolute knowledge;
- *Relatedness*: degree to which a knowledge seeker is familiar or has prior experience with a given knowledge platform, principles or context;
- *Uniqueness*: presence or absence of alternate, substitutable bodies of knowledge;
- *Value*: absolute (market value) or relative (value to a partner given its specific capabilities, history, context and ambitions);
- *Actionability*: readiness, receptivity and ability of the organisation to use and control particular knowledge.

Similar to the discussion of objective and constructed knowledge in learning theories, knowledge can also be understood from the perspective of the relevance of its context (Gherardi 1999; Swan 2003), the latter in the sense of 'a set of alternatives made of constraints and enablements, within which individual (or collective) actors *can* or *have* to choose' (Kazepov 2004: 6). *Context-independent knowledge* is an object that does not need a certain context to be valid and understood, but that can be codified, stored and transferred from one setting and person to another and thus de-contextualised. This is mainly the case for knowledge that is scientifically justified and universally applicable (such as in natural sciences). *Context-dependent knowledge* is situated in people's heads, a result of social interactions and limited in terms of codification (Maaninen-Olson et al. 2008). This perspective acknowledges the limits of knowledge codification.

3.3 Cooperation

An increasing number of companies and public institutions are looking for diverse and complementary resources and knowledge that are not available internally, particularly in the field of innovation. In this context, learning from the experience of other organisations becomes relevant, especially if partners possess somewhat different experience and capabilities. Different experiences might arise from working in different geographical, cultural and political environments, and different capabilities might arise from working in different areas such as research, production, and government. Cooperation can take many forms, such as non-profit and profit-non-profit collaborations as well as public sector alliances, but relevant research has mainly focused on the business world.

Many scholars assume that the prime objective of collaborative alliances is knowledge acquisition and learning that can speed up the diffusion of innovation at lower costs (for example Mamadouh et al. 2002; Khanna et al. 1998; Mitchell and Singh 1996). Beyond the access to the knowledge of others, partners may also be motivated to 'teach' rather than to 'learn', meaning that the knowledge holder is interested in diffusing their existing knowledge and practices to cooperation partners. *Learning from each other* allows access to partners' skills and knowledge, but cooperation also permits the creation of new knowledge. *Learning with each other* ('generative learning') refers to the development of new knowledge, including the management of cooperation or of cultural differences and unfamiliar environments. It is also understood that cooperation eases the persuasion of those reluctant to change by pointing to successful results in other countries.

The main challenges that cooperation faces in international, inter-organisational and inter-disciplinary settings as well as in project environments are discussed in the following sections.

3.3.1 Collaborative Learning in International, Inter-organisational and Inter-disciplinary Environments

Transnational projects where participating actors are from different backgrounds and settings can have a high *variety of identities and contexts*. Consequently, inter-organisational relations have to deal with a greater degree of complexity and dynamics than intra-organisational learning processes. In particular the international environment creates barriers to effective knowledge transfer as 'internationality . . . adds a new layer of complexity to the tasks of creating, transferring, applying, and exploiting knowledge' (Macharzina et al. 2001: 632). This can affect both international knowledge generation and the ability to use knowledge internationally. Transnational INTERREG projects are particularly challenged with context and thus cultural aspects as they are asked to consider integrated regional perspectives. Many of them, therefore, deal with issues related to the planning discipline, which again has to deal with major differences in planning culture, that is, of its conception, institutionalisation and implementation (Friedman 2005). This is much less the case in disciplines such as civil engineering, as will be seen in one of the case studies. Although transnational cooperation is supported as part of the European integration project, research on different planning cultures shows that neither globalisation nor Europeanisation has led to the homogenisation of planning cultures (Friedman 2005; Sanyal 2005; Dühr and Nadin 2007).

Problems related to knowledge transfer can arise due to strategic, structural, political-cultural and individual barriers (Prange et al. 1996; Schüppel 1997). Ideas that are new to one organisation can open new perspectives but at the same time need to be compatible with its existing practices (van Bueren et al. 2002). This requires the recipient to have a reflexive relationship towards learning. *Language boundaries* (for example functional, organisational, national) add another layer of complexity. *Cultural distance* based on national or organisational cultures may enhance the usefulness of knowledge transfer but can also increase its cost while decreasing its comprehension and speed (Pérez-Nordtvedt et al. 2008).

Whether *diversity* is an asset or a barrier for cooperation may also depend on the tasks to perform and on the nature of the diversity. Hambrick et al. (1998) found that in cases of creative tasks, different values (fundamental preferences) have a positive effect related to group effectiveness, but a negative effect in cases of coordinative tasks (involving elaborate interaction among group members). In addition, disparate values can create interpersonal strains and mistrust. In contrast, cognitive diversity (knowledge, assumptions, schema) was found to enhance the effectiveness of creative, coordinative and computational tasks (analysis of clear-cut information based on relatively objective standards), but only up to the point at which knowledge is explicitly required for the task. Beyond that, cognitive diversity may even become counter-productive for coordinative tasks. The authors found that diversity in behaviour decreases the effectiveness for coordinative tasks as it decreases the 'ease of communication'. Diversity management can help to overcome these effects (Hachmann and Potter 2007).

Transnational projects in the INTERREG programmes are supposed to follow an 'integrated approach' and be of 'territorial relevance' (Interreg North-West Europe 2015), which means that challenges are not only to be addressed by technical solutions but also by societal, governance or communicative measures. Consequently, partnerships are often characterised by an *interdisciplinary* nature. INTERREG programmes require projects to pursue an integrated approach, but the complexity of many of the addressed issues simply calls for an interdisciplinary approach, as they are not isolated to particular sectors as many territorial challenges picked up by INTERREG projects fall into the domain of different disciplines (Thompson Klein 2004). However, the interdisciplinary nature of projects adds even more complexity to the collaboration. Although studies on interdisciplinary research are rare, it is known that in projects interdisciplinarity increases the difficulty to find suitable cooperation partners and to establish fruitful collaborative patterns as relevant actors are from different professional worlds (Bruce et al. 2004). How far the involved disciplines collaborate or work side by side differs between projects. In a study on cooperation in the EU's Research Framework Programme, the authors show that there were more 'multidisciplinary projects' (with a range of disciplines that work in a self-contained manner with little cross-fertilisation) than 'interdisciplinary projects' (where the contributions of the various disciplines are integrated to provide a holistic and systemic outcome) (ibid.).

In case of different systems of meaning between the sender and the recipient of knowledge, the cause-and-effect relationships between the transferred knowledge and the expected effects can be distorted and knowledge may have to be reconfigured to fit the interpretative repertoire of the recipient (Huelsmann et al. 2005; Macharzina et al. 2001). Thus, a challenge lies in the translation of implicit knowledge embedded in the context into explicit, codifiable knowledge by de-contextualisation.

3.3.2 Knowledge Development and Learning in Projects

In the past, there has been a shift in various disciplines in terms of the mode of knowledge production. The interest has moved from the traditional science-based institutional framework to knowledge production in the context of its application, increasingly involving inter- and transdisciplinarity and diversity. This has led to the development of research on 'temporary organisations' as knowledge production increasingly takes place in temporary organisational settings that include a multiplicity of sources. However, most studies have focused on intra-organisational temporary organisations while the number of studies dealing with inter-organisational relationships is still scarce.

Janowicz-Panjaitan et al. (2009: 58) describe temporary organisations as 'set up to accomplish one or a very limited number of tasks, and to do so through a team of selected actors within a limited amount of time and with transition as an ultimate end'. In addition, they tackle 'tasks of higher complexity' and engender 'higher uncertainty and interdependence between team members while simultaneously having more time and budget constraints' compared to permanent organisations. Temporary organisations are temporary work teams and networks, which disperse when a problem is solved or redefined. Members then move to different groups involving different people and focus on different problems. As temporary work practice does not have the same supporting structures and routines as permanent work practice and as there is a strong focus on project goals and tasks, knowledge is more individualised and learning is primarily local (Maaninen-Olsson et al. 2008). Later on, this affects the integration of knowledge in the relevant home organisations.

Some authors have developed frameworks for researching learning in temporary project-based organisations and distinguishing them from permanent organisations (see section 4.1). These provide analytical variables of alliance- and strategy-specific characteristics of transnational projects.

Within the research field of temporary organisations, the strand of '*project-based organisations*' focuses on temporary organisations based on projects. This goes well beyond traditional project management literature, which is usually highly normative, describes projects as 'tools' rather than 'organisations' and neglects the role of individuals and especially of their motivations to participate. However, the complexity of individual motivations, cultures and conceptions often turns projects into difficult to steer and influence entities that are not always 'exciting, non-hierarchical, and stimulating experiences, in which the team spirit can flourish and creativeness be nourished' (Packendorff 1995: 326), as they are often referred to in traditional project management literature. Studying projects as action systems, Packendorff follows, 'means putting less energy into studying what is meant to happen, and more into what is actually happening' (ibid: 330).

In general, the shift to an understanding of projects as temporary organisations has led to the focus switching from questions of project planning and structures towards *organising processes*, that is 'the deliberate social interaction occurring between people working together to accomplish a certain, inter-subjectively determined task' (ibid: 328). Although planning and structures are still important inputs into such processes, it is the inter-subjective meaning given to plans and structural arrangements by the participants that 'explains' whatever action is taken with reference to these phenomena (ibid.). As Söderholm (1991) emphasises, the relationship between the acting individual and the structure can thus be 'translated' into a relationship between process and structure, since the process concept represents the actions of a number of persons.

In the field of project-based learning, the question of knowledge transferability has been taken up by Carlile (2002, 2004), who identified different 'knowledge boundaries'. These can be of syntactic or information-processing nature (differences between sender and receiver), of semantic or interpretative nature (unclear differences and dependencies, ambiguous meanings) or of pragmatic or political nature (diverse interests need to be negotiated). The fact that knowledge is localised around problems faced by a given practice, embedded in the experiences, know-how, technologies and methods used in that given practice and incorporated in methods and applications may become problematic when working across practices (ibid). The analysis of a cross-border project shows in which way these boundaries appear in relation to territorial issues (differences in language, culture and institutional context), roles (differences in interests, means and roles), sectors and project involvement (Valkering et al. 2013).

3.4 Conclusion

Learning

In psychological studies, the different learning theories developed in an evolutionary process and can be seen as complementing each other. Social-cognitive learning theory and the concept of cultural influence on people's perception of reality contribute to a better understanding of learning processes in projects and transnational cooperation. Both social-cognitive and constructivist learning theories also emphasise the role of prior experience.

Similar to learning theories in psychological studies, concepts of organisational learning developed in an evolutionary process. Applied to an organisational context, behaviouristic approaches focus on studying observable changes in practices, while cognitive approaches explain the development of individual and group learning processes and social aspects. Organisational learning mainly deals with learning at the level of individual private businesses and less with group processes. Group learning takes place as a consequence of project learning and – beyond individual learning effects – can be linked to project achievements. As mentioned before, this book is not concerned with evaluating how single organisations learn from transnational projects, but how projects learn in transnational settings. In general, organisational learning theory lacks procedural aspects and its contribution to conceptualising transnational learning processes is thus limited.

Still, the concept of different levels of learning developed in organisational learning literature supports a focus on the group and project level and the distinction from processes at the level of individuals and organisations. The latter is, of course, paramount for implementing project results and ensuring their long-term survival. With respect to transnational learning, the most relevant analytical levels are the sharing of experience and ideas between individual partners, the collective incorporation of knowledge into collective capacity building processes and, based on this common starting point, collective decisions and initiative. Moreover, the idea of different levels of learning contributes to the understanding of the degree and depth at which learning takes place and to their relationship to different types of learning effects.

Knowledge

Different categories and typologies of knowledge assist the analysis of project subjects. In the context of practitioners with varying professional, institutional and cultural backgrounds cooperating in transnational projects in highly diverse environments, the conceptualisation of knowledge into tacit and explicit knowledge, its potential context-dependency, as well as different degrees of complexity, relatedness, and novelty determine the challenges of joint learning processes. They all add to the appreciation of different levels of knowledge transferability, which impacts on the success of inter-partner knowledge transfer.

In the context of European Territorial Cooperation, knowledge development is challenged by a variety of factors related to its transnational, inter-organisational and inter-disciplinary character. The international and inter-organisational character of projects accounts for cultural and contextual diversity as well as language barriers. Diversity has been found to have different effects, depending on the type of performed tasks and the type of diversity concerned. The requirement for strategic and integrative project approaches adds challenges related to interdisciplinary work, notably complexity and different approaches, logics and languages.

Cooperation

The literature on 'temporary organisations' and 'project-based organisations' allows the understanding of cooperation projects with respect to two of their key features – that is, their temporariness and project character. It emphasises the relevance of studying processes when researching projects; that is the inter-subjective meaning given to structural project arrangements. In general, knowledge development and learning in temporary project-based organisations proceed differently and face different challenges than in non-temporary organisations. 'Project knowledge' is a resource for action in a project that can be found within the project's result and that is subject to changes during the process of the project's execution.

This literature strand contributes framework conditions for learning in projects but is otherwise mostly concerned with the integration of project-produced knowledge in the relevant home organisations rather than with knowledge production itself. It stresses that knowledge development in temporary work practice is more individualised and primarily local due to a strong focus on project goals and tasks and a lack of supporting structures and routines. The main challenge of knowledge development in projects is the difficulty for organisational learning due to much of the knowledge generated in the project activities being embedded in tacit experiences of the group members and the risk of knowledge being dispersed as soon as the project is dissolved. 'Knowledge boundaries' limit the transfer of project knowledge between project participants and between the project and the participants' home organisations. Knowledge development in projects thus faces the 'paradox of project-based learning': while individual projects offer an environment conducive to learning, they may also create strong barriers to the continuity of learning beyond project boundaries.

4   Towards a Model for Transnational Knowledge Development and Learning

This chapter develops an analytical model for the conceptualisation of knowledge development and transfer in transnational projects that builds on contributions from a variety of theory strands including (inter-) organisational studies, educational studies and policy transfer. In a first step, a structural framework is drawn up that encompasses relevant parameters that shape the conditions under which projects take place. As these can differ considerably between transnational projects, cooperation processes can be of highly different character. Particularly, the three sub-fields of organisational studies, inter-organisational learning, temporary organisations and project-based organisations, provide valuable insights into cooperation structures and structural variables applicable to collaborative projects. In a second step, the perspective is shifted from project structures to processes. With the help of models of experiential learning from educational studies and models for knowledge creation from organisational studies, an attempt is made to conceptualise the main process phases that transnational projects pass through. The focus is on those phases that take place during project lifetime and are part of the project scope that is learning at project level. This excludes the steps taken by organisations to implement project findings as a result of cooperation, which, arguably, are highly decisive for long-term learning at organisational and regional levels, but which rely on learning achievements made at project level in the first place.

Figure 4.1 summarises the interaction between project structures and cooperation processes in a hypothetical model that forms the holistic basis for the development of a more detailed analytical model for transnational knowledge development and learning. Programme-funded transnational projects are embedded in a given programme framework. In the case of the INTERREG programmes, this consists of both the general EU framework of relevant guidelines and the specific Operational Programme of the individual cooperation area (see section 2.1). Within this programme framework, individual, and potentially interlinked, structural parameters determine the project framework. These form the starting point for cooperation processes as well as knowledge and learning processes. In terms of learning, the process perspective describes the transformation of individual knowledge input into new transnational knowledge through sharing, transferring, processing and advancement.

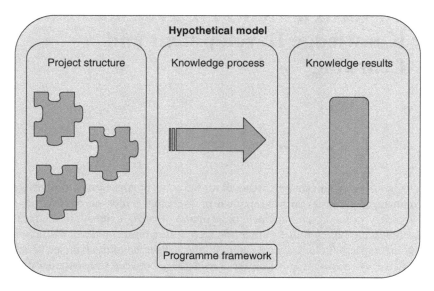

Figure 4.1 Hypothetical model of project structures and processes

Source: by author

This perspective approaches transnational projects as open, temporary, complex and dynamic systems. It is a conceptual framework for transnational knowledge development and learning and consists of constructs, parameters and interdependencies relevant to understand, describe and explain how knowledge is developed in transnational projects.

4.1 Project Properties: Setting the Scene for Transnational Knowledge Creation and Learning

All project processes are based on the given structures that shape their frameworks, such as a certain time perspective and a specific scope of partners and objectives. Funding programmes usually pre-define and limit some of these structures, such as a minimum or maximum project lifetime and financial scope, a minimum or maximum amount of cooperation partners or eligible project topics. Within this – often strongly quantitative – default framework, projects can still choose from a breadth of qualitative factors and thereby prepare the ground for very different types of cooperation processes. Findings from both inter-organisational learning studies and research in the field of temporary and project-based organisations provide theoretical models to analyse the properties and preconditions of collaborative projects. They differ in the way they link structural factors to cooperation processes, but all provide at least some ideas as to which structures can be related to certain processes and specific challenges to cooperation.

Theoretical structural frameworks for researching knowledge development and transfer in alliances and projects can be traced back to three groups of authors. In the field of inter-organisational learning, Salk and Simonin (2008) created a framework that is made up of four blocks of variables: (I) partner-specific, (II) alliance-specific, (III) knowledge-specific, and (IV) context-specific variables. In the field of temporary project-based organisations, Dietrich et al. (2007) and Lundin and Söderblom (1995) set up comparable models. While the former identify actors, purpose, context and action, the latter work with time, task, team and transition. These variables are summarised in Figure 4.2 and used as the basis of the analytical framework of this book.

With respect to project partners, projects are dependent on the will, commitment, prior experiences and skills of the people involved in their creation, development and termination. Other relevant *partner-related variables* involve collaborative know-how, partners' intents, trust and culture as well as partners' 'absorptive capacity' (see box below). Moreover, Dietrich et al. (2007) add the actors' involvement in the project's tasks and their relationships with external stakeholders.

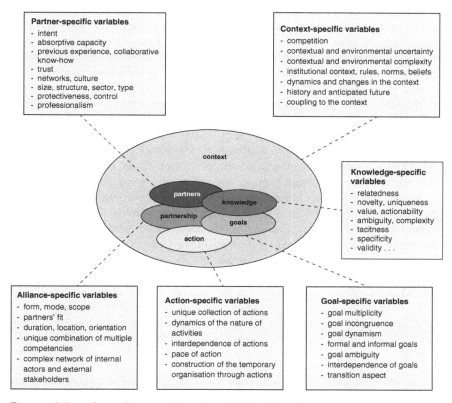

Figure 4.2 Learning and partnerships: theoretical building blocks

Source: adapted from Salk and Simonin 2008: 265; Dietrich et al. 2007: 9; Lundin and Söderblom 1995: 439

Alliance-specific variables relate to aspects such as organisational form, the 'fit' of the involved partners (that is the unique combination of partners in a project), the duration of the alliance or its location. When it comes to the scope of alliance activities, the type of activity, their relatedness to the partners' expertise and core competencies and their general novelty can be distinguished.

Knowledge-specific variables refer to knowledge characteristics (see Chapter 3) that determine for example the ambiguity and value of knowledge. The ambiguity of knowledge can impact on the comprehension and transferability of knowledge while the value of knowledge can stimulate the learning motivation of the knowledge seeker.

The *context* of a project refers to the fact that projects are related to previous and simultaneous activities, plans, standard operating procedures, traditions and norms in their organisational contexts. These variables are mainly non-controllable factors determined by the environment in which the collaboration takes place.

With respect to a project's *purpose*, Dietrich et al. (2007) bring together various research findings that show how goals can relate to different time frames, different levels of the project, different levels of the parent organisation and different levels of abstraction. They may be loosely or tightly coupled to the parent organisation and there can be interdependencies between the goals of the project and those of its parent organisations. The success of projects is often judged against the achievement of predetermined objectives (Packendorff 1995), but as this can be a matter of interpretation, it may be very difficult to determine. Instead of being predetermined targets, goals may sometimes be more like summaries of actions (Dietrich et al. 2007). The project purpose usually contains an element of change, a qualitative difference between the 'before' and 'after' (Lundin and Söderholm 1995). As will be seen later, this was the case in some of the transnational case studies.

In terms of project *action*, Dietrich et al. argue with Lundin and Söderholm (1995) that although organisational theories often focus on decision-making as a predominant factor that explains the nature of organisations, the theory of temporary organisations is based on the notion of action rather than decision. Projects are built around the need to perform certain actions. Different tasks are interrelated to each other on various levels, such as management actions that are interrelated to content actions. Action in a temporary organisation particularly differs from that in permanent organisations due to a strong emphasis on time schedules and deadlines, but also due to a unique collection of actions. Projects depend on the dynamics of the nature of activities, their interdependencies and pace.

To the theoretical frameworks presented above, Lubatkin et al. (2001) – with their framework of cooperative learning theory based on educational and social psychology – add three major interdependencies necessary for cooperative learning processes:

- resource interdependence guarantees that every participant has some knowledge that is 'unique' and not possessed by the others;
- with goal interdependence participants realise that they can best achieve their goals by cooperation;

- with task interdependence participants understand that their agenda is best realised when they specialise in the activities at which they are individually most competent.

The concept of absorptive capacity

In knowledge transfer, the concept of 'absorptive capacity' calls particular attention to the sender's and the recipient's ability to recognise the value of knowledge and to make use of it (Cohen and Levinthal 1990). It links intra- and inter-organisational knowledge transfer and learning. Once knowledge has entered the recipient's sphere, its distribution within the organisation depends on its intra-organisational knowledge transfer capability. If the received knowledge is too different from an organisation's mental representation, it may easily be ignored or treated as something unique and therefore not taken seriously. At the same time, the sender needs absorptive capacity to judge what knowledge could be useful for the recipient and intra-organisational transfer capability to make the knowledge accessible in an efficient manner.

Lubatkin et al. (2001) propose that – similar to 'absorptive capacity' – there is a 'reciprocal learning capacity' with respect to joint knowledge development. Based on the educational and social psychological research tradition, they develop a list of requirements for successful learning:

- a similar general knowledge base, or put simply, 'there can be no coexperimentation with abstract knowledge without both partners having the ability to speak the basics of each other's language';
- different areas of expertise, or in other words, 'to learn implies that something new is taught';
- similar organisational values and routines, or 'organisational fit' that facilitates learning and reduces the threats of misunderstandings as partners recognise and appreciate each other's abstract and tacit know-hows;
- a similar unifying vision and strategic motivations;
- a positive reputation of being a 'good' partner and cooperator.

Pérez-Nordtvedt et al. (2008) add:

- the recipient's learning intention (the desire to learn from another entity);
- the attractiveness of the source (for example trustworthiness, country of origin, perceived reliability, local embeddedness, strategic importance, and relationship quality).

Following this view, it could be assumed that organisations looking to explore and exploit new knowledge would choose their learning partner according to at least some of these aspects. A range of studies on project learning show that a high level of absorptive capacity is a necessary ingredient for successful knowledge transfer, but also that it needs to be combined with cognitive proximity (such as a 'common language') and cooperation (Bakker et al. 2011).

Table 4.1 Parameters for knowledge transfer and learning processes

Parameter group	Parameters used in the analysis
Partner-specific characteristics	Organisational type (including structure, culture, practices), motivation for participation, previous experience, identified knowledge gaps and potential
Partnership-specific characteristics	General set-up given by form and mode, stage; partners' fit (including different/similar knowledge bases), extent of common knowledge, relationship between sender-receiver, diversity, strategic fit, geographical scope
Knowledge-specific characteristics	Types of knowledge identified by Salk and Simonin 2008, knowledge-orientation of project
Strategy-specific characteristics	Project tasks and objectives, tasks, division of labour and integration, character and fit of pilot projects
Process-related characteristics	Communication, intensity of cooperation, scope for interaction, knowledge sources, transferability, exchange, mechanisms for knowledge transfer and learning, reflection and feedback, knowledge-processing

Source: by author

For the analytical model, parameters unique to the project context and individual partner characteristics were not taken up, as they would have required different research approaches (for example interviews with additional colleagues, psychological questions) or gone beyond organisational aspects of projects (for example partners' political framework, spatial construction and identity). Particularly individual aspects often lead into subjective areas and require a certain degree of self-awareness among interviewees. Due to similar reasons, processes of group socialisation and other 'soft' factors (for example power, trust) were not included. However, it needs to be kept in mind that they can contribute considerably to project dynamics (Sol et al. 2013; Newig et al. 2010; Easterby-Smith et al. 2008). Table 4.1 summarises the parameters chosen as analytical units in the analytical model.

4.2   The Process Perspective: Opening the Black Box of Transnational Knowledge Creation and Learning

Transnational projects cooperate and exchange and develop knowledge based on their given project structures. Project processes are thus path-dependent. In the following, several process models for learning are discussed that help to open up the 'black box' of transnational project processes. Experiential learning models contribute notions of 'making experience' that changes existing knowledge and adds new knowledge by making use of feedback loops. This is highly relevant for practical projects that often include the testing and demonstration of concepts, strategies and findings in 'pilot projects'. Still, experiential learning usually only focuses on individual learning. Knowledge creation models enrich the understanding of learning processes by adding social aspects to knowledge development, concepts of relevant phases groups of learners pass through and learning mechanisms that help these processes.

4.2.1 Experiential Learning Models

Experiences play a vital role when it comes to learning; in a way, one could argue that they feed all learning processes. When people learn from experience, they use their knowledge in a reflexive way when dealing with a particular case. Experience is particularly relevant in a practical project context and introduces the notion of path-dependency of learning processes.

Cyclical models that explain learning by experience developed since the 1990s go back to the experiential learning model by the educational theorist Kolb (1984), who based his work on earlier psychological theorists such as Piaget (1970), Lewin (1951) or Dewey (1938). In Kolb's model, learning is the process whereby knowledge is created through the transformation of experience. Learning is a series of transitions of four adaptive modes:

- concrete experience;
- reflective observation;
- abstract conceptualisation; and
- active experimentation.

Learners translate their experience through observation and reflection into more or less abstract concepts, which can be used as guides for new situations. This process requires learners to reflect on their experiences before they can make meaning of them and then progress to adapt their behaviour based on what they learned. Modification of knowledge can occur in cases where the results of an action do not correspond to the experience of that action. Then people reflect on their generalised knowledge as well as on the specific experience, question and adapt underlying linkages (Humpl 2004). In addition to skills for reflection, learners require an open and unprejudiced engagement in new experience, observation skills, and the ability to create concepts that integrate observations into theories and use these theories for decision making and problem solving (Kolb 1984).

With respect to learning processes, two aspects of experiences are important (Humpl 2004): firstly, as experiences are contextual, one of the largest intellectual challenges is to apply them to new situations. Secondly, not everything that is perceived is necessarily considered an experience: experiences rather evolve in the discrepancy between expectations and perception. If perceptions of the environment are affirmed, no 'new' experiences are made. Instead, expectations are still identical with perceptions.

In Kolb's model, this unit remains ambiguous but has been used to explain individual, team and organisational learning. The model does not take into account that in coping with a concrete experience, people rely on previous experience and that this makes an open and unprejudiced engagement in experience in practice impossible. In case of similar experiences, generalisations or analogies can occur (Humpl 2004). Although they usually focus on individual learning, experiential learning models are interesting in the context of transnational cooperation projects as they add the concept of 'experience' to knowledge acquisition. In projects that mainly involve practitioners such as INTERREG projects, large parts of the potential learning process are based on practical experience, particularly in the so-called 'pilot projects' (see section 4.3.1). Experiential learning models can shed light on how practical project experience serves as a basis for knowledge development and learning, for example on the role of reflective practice. However, they do not consider how joint reflection processes take place – especially when different mind frames are involved – or how joint learning can build on very different experiences, such as in diverse pilot projects.

4.2.2 Knowledge Creation Models

Literature of organisational learning is largely based on individual learning processes in organisations, on their effects on the organisation and rather static considerations of the conditions for successful organisational learning. Social interactions in and between organisations that lead to learning as well as the process of learning as such remain a 'black box'. Still, knowledge creation models include social and/or procedural aspects of learning processes and thus go beyond simple input-output schemes. The oldest of these is the model of different levels in organisational learning by Argyris and Schön (see section 3.1.2). Of later models that include both social and procedural aspects, the knowledge-creation cycle developed by Nonaka and Takeuchi (1995) has been most influential. Their SECI model includes a perspective on knowledge qualities (tacit and explicit) and thus discusses the relationship between the learning process and learning contents and – at least partly – embraces the idea of 'experiential knowledge'. It adds social aspects when conceptualising knowledge development through different stages of group interaction. Models of learning typologies and mechanisms pick up the concept of different learning phases and shed light on which learning approaches and tools are useful in the different learning phases (for example Prencipe and Tell 2001). Finally, the model for inter-organisational knowledge creation by Huelsmann et al. (2005) also focuses on its social dimension and adds the dimension of (I) *learning interactions* either between single individuals (knowledge transfer) or at group-wide level (knowledge sharing) and of (II) the *formation date of knowledge* to distinguish between the movement of existing knowledge and the joint development of new knowledge.

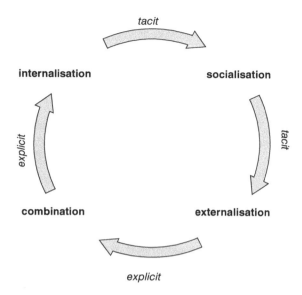

Figure 4.3 Knowledge Spiral by Nonaka and Takeuchi: SECI model

Source: adapted from Nonaka 1994

A The SECI model

The SECI model is based on the critique of conventional organisational learn-ing literature that mostly focuses on processes of individual learning and passive organisational learning that adapts to external changes and fails to address knowledge-creation resulting from learning (Nonaka et al. 2001). In their dynamic model (see Figure 4.3), the authors address the question of how knowledge develops in a social process and how it is converted into concepts, models and structures that can change organisational behaviour. Their observations are based on knowledge creation at the individual, team and organisational level.

The model uses Polanyi´s distinction between tacit and explicit knowledge, but in a slightly different understanding. For Polanyi, tacit knowledge is the background for explicit knowledge and the two are not seen separately. The SECI model, on the other hand, contrasts tacit knowledge with articulated knowledge. Particularly the translation of tacit into explicit knowledge thus needs to be treated with care.

1   In a first step, sharing experience transfers tacit knowledge, a process the authors call *socialisation*. This takes place in a 'space' of social interaction by observation, imitation and practice.
2   In a second step, tacit knowledge is articulated into explicit concepts by dialogue, collective reflecting and concept building, the so-called *externalisation*.

Dialogue and collective reflection trigger this process and it is supported by reasoning and communication methods (for example deduction, induction, dialectical reasoning, metaphors, contradictions, analogies, iterative processes, figurative language).

3   In the following *combination* mode, activities such as sorting, adding and categorising combine explicit knowledge and thereby produce new knowledge. Networking and integrating new knowledge into existing stocks of knowledge are the basis for the combination mode.

4   Finally, explicit knowledge is re-embodied into tacit knowledge in the form of increased individual skills and competencies (*internalisation*); including both the transmission and application of tacit knowledge to activities. This step corresponds to individual learning and is supported by 'learning by doing'.

In all steps, communication plays a significant role and, therefore, the underlying information and communication structures support the whole process of knowledge creation.

Interestingly, the SECI model combines a social view of learning with an individualistic conception of knowledge. While knowledge is collectively created, it later needs to be internalised at the individual level, which means that while the process is social, the result is not (Tuomi 1999). In cases of different kinds of organisation cultures, language and general cultural backgrounds, further factors have to be taken into account (ibid.). When applying the SECI model to INTERREG projects, the projects allow socialisation through the general cooperation process, externalisation through discussions and combination mainly through report writing. The final internalisation of knowledge lies outside the project domain and in the responsibility of individual participants and their home organisations.

B Learning Processes, Typologies and Mechanisms

The model of learning typologies by Prencipe and Tell (2001) uses three phases of learning processes originally identified by Zollo and Winter (2001): experience accumulation, knowledge articulation and knowledge codification. Prencipe and Tell analyse these processes with respect to dominant learning typologies, potential outcomes and economic benefits.

• In *experience accumulation*, they argue, actors are likely to learn locally and closely related to existing routines. Learning by doing and learning by using are both based on experience from actions and related to single-loop learning.

• *Knowledge articulation* through reflection, thinking, discussing and confronting allows understanding causality and feasibility and can have a collective element when communicated. This can create an arena for double-loop learning.

• *Codification*, finally, can be seen as an extension of articulation and allows the creation of externalised knowledge. Creative elements such as the screening of multiple scenarios, different explanatory frameworks or the testing of various organising principles can help in this respect.

These learning typologies are directly relevant for learning in INTERREG projects: experience accumulation takes place in pilot projects and other activities, knowledge articulation through transnational discussions and reflection and knowledge codification through report writing and implementation.

C Models for Inter-organisational Knowledge Creation

In their theoretical framework for inter-organisational learning based on a system-theoretical approach, Huelsmann et al. (2005) focus on the social dimension of learning. They distinguish between three different perspectives: (I) learning interactions that are limited to single individuals transferring existing knowledge, (II) learning interactions that share existing knowledge at a wider group level knowledge, and (III) the joint development of new knowledge.

Knowledge Transfer: This perspective refers to learning from other organisations, focusing on learning processes and relations between partners. Here, inter-organisational learning deals with the movement of existing knowledge, technology or practice, for which such transfer represents a knowledge input and is related to internal knowledge processing.

Knowledge Sharing: This perspective allows the whole cooperation project to achieve a benefit by sharing knowledge. The inter-organisational knowledge base permits an increased variety of knowledge as well as synergy effects.

Joint Knowledge Development: The third perspective, the development of new knowledge, has been only little researched so far. Here, new knowledge is created due to synergy effects within the shared, inter-organisational knowledge base. In these 'reciprocal learning alliances' (Lubatkin et al. 2001), the key intent of the involved partners is to co-experiment and to make use of each other's unique, but complementary knowledge structures, to bring together all information domains and create new knowledge.

Although INTERREG projects allow knowledge transfer between single project partners, their multilateral settings are primarily appropriate for knowledge sharing. In transnational settings, knowledge transfer is particularly faced with challenges related to the limited transferability of knowledge (see section 4.3.2). Joint knowledge development again is of importance in projects that do not only aim at spreading innovation but at innovating (explorative projects). Moreover, all projects that aim at creating joint results and products usually require joint knowledge development to have taken place.

4.3 Phases of Learning in Transnational Cooperation Projects

Based on the discussed models of learning processes, the following is an attempt to conceptualise the main process phases of transnational projects on their way to produce their projected results in a transnational approach. The discussed models of learning processes vary with respect to where and when learning starts, that is when an experience is made (Kolb, Prencipe and Tell), or when this experience is shared by a group or an organisation (Nonaka and Takeuchi; Huelsmann at al.).

When looking at the different models from an integrated perspective and applying them to the reality of transnational cooperation projects, six major process phases can be identified:

1   Making experience ('concrete experience', 'experience accumulation');
2   Exchanging experience ('socialisation', 'knowledge transfer');
3   Reflecting ('reflective observation', 'externalisation', 'knowledge articulation', 'knowledge sharing');
4   Abstracting and conceptualising ('abstract conceptualisation', 'combination', 'knowledge codification', 'joint knowledge development');
5   Internalisation of knowledge ('internalisation');
6   Adoption of behaviour and application of learning results ('consolidating new practice').

As explained above, the model focuses on phases of direct project processes and of the immediate project scope. Steps 5 and 6 are usually not part of the project scope or directly part of project results, but are mid- and long-term effects. In other words, while the first four phases are phases of project learning, the individual and organisational internalisation and implementation of lessons learned are steps, albeit highly relevant, beyond direct transnational concern and are thus not followed up further.

The first four process phases can be linked to typical project elements of transnational cooperation projects as well as to relevant skills identified in the above-discussed models and summarised in Table 4.2.

The making of experience is relevant in transnational projects that include testing and demonstration activities in so-called 'pilot projects'. Overall, transnational projects are focused on the exchange of experiences and knowledge between project partners, which usually takes place during regular project meetings and study visits as well as in work groups or common actions. Reflection processes can occur anytime during project discussions but are often linked to targeted discussions in work groups, for example in relation to project 'work packages'. The diversity of partners and the abundance of individual experiences require a certain systematisation and storing of knowledge in order to make sense together, but projects also face the challenge of translating individual experiential knowledge into common and more explicit knowledge. Thus, abstraction is required for the combination and composition of explicit knowledge to produce the project results, which often take the form of written products such as guidebooks or final reports.

The following section looks into the four phases in more detail as well as common related barriers.

4.3.1 Phase 1: Making Experiences

A particular feature of transnational projects in the INTERREG framework as well as in many other transnational programmes is the concept of 'pilot projects'

Table 4.2 Transnational project elements in the identified process phases

Learning phases	Typical elements of transnational cooperation projects	Skills required
1 **Making experience**	Pilot projects	Openness, impartiality, engagement in making experience
2 **Exchanging experience and knowledge** A of new experience B Transfer of existing knowledge	Transnational meetings, study visits, work groups, joint actions	Creating a 'space', observation skills
3 **Reflection**	Transnational discussions, work groups	Reflection skills, dialogue, cooperation
4 **Abstraction**	Composing the final results and products (guidebooks, tools, final report, etc.)	Networking, integrating knowledge into existing stocks of knowledge, writing, creating concepts that integrate observation into theories

Source: by author

or 'transnational investments'. While working on joint objectives and following a common scheme of work packages, partners often implement actions at the local level and thereby make new experiences related to the overall project topic and its cognitive interest. Pilot projects are supposed to work as test cases for innovative solutions developed by project partners in a confined setting and allow the studying of the interaction between the solution and its context. They can work as 'niches', protected environments, in which innovations are developed and applied (van Mierlo 2012). Lessons learned can then be used to improve the solution or to adjust management practices and policies (Vreugdenhill et al. 2010). Pilot projects can take different forms and pursue various purposes, such as exploration (research-driven), communication (management-driven) or advocacy (political-entrepreneurial) (ibid.).

4.3.2 Phase 2: Exchanging Knowledge and Experience

The dominant process in transnational cooperation projects is the sharing of experience and knowledge between participating partners. In terms of the source of relevant experience and knowledge, a distinction can be made between existing knowledge stocks and newly made experience. As these are of very different quality, in the following they are treated as two distinct elements. While new experience from pilot projects mainly relates to 'raw' experience that needs to be reflected upon during the course of the project in order to be used, exchanging existing stocks of knowledge is related to knowledge transfer processes that involve 'senders' and 'receivers' of knowledge.

In the following, insights from policy transfer literature are discussed to better understand the relevant processes of knowledge transfer from existing knowledge stocks. Although most of the policy transfer literature concentrates on transfer between governments in bilateral learning settings and on the receiving side of information, it contributes useful categorisations of transfer objects and degrees of transfer applicable to other contexts. Moreover, findings on the conditions that influence the success of knowledge transfer and especially conceptions of different degrees of knowledge transferability are useful for the understanding of knowledge transfer and sharing in transnational cooperation.

This is followed by a discussion of the potential provided by the parallel implementation of individual pilot projects by project partners. This allows partners to experience not only their pilot project in a familiar environment but also those of their transnational partners, which may allow studying similar projects in very different social-economic frameworks. The concept of 'observational learning' developed by Bandura (1979) explains how learning impulses can originate both in direct and indirect experience and under what conditions a portfolio of pilot projects contributes most effectively to transnational learning.

A   Knowledge Transfer from Existing Knowledge Stocks

Project partners with their different knowledge and experience bases that relate to diverse societal, legal and economic frameworks are rich sources of knowledge and inspiration in transnational projects. Transfer of knowledge can take place at any time during project lifetime and contributes to a more informed and effective planning and implementation of the whole project, including its pilot projects and joint actions.

In general, knowledge transfer between organisations describes an event through which one organisation learns from the experience of another. In recent years, various empirical studies on knowledge transfer have been carried out in a variety of governance arrangements (for example Becker and Præst Knudsen 2006). The discipline of political science and comparative politics has particularly well-developed conceptualisations of knowledge transfer, in terms of 'policy transfer'. Policy transfer describes processes of governmental learning by which 'knowledge about how policies, administrative arrangements, institutions and ideas in one political setting (past or present) is used in the development of policies, administrative arrangements, institutions and ideas in another political setting' (Dolowitz and Marsh 2000: 5). Lesson drawing (James and Lodge 2003; Rose 1993; Robertson 1991), policy learning (Bennett and Howlett 1992), policy transfer (Benson and Jordan 2012; Savi and Randma-Liv 2013; De Jong and Edelenbos 2007; Evans 2004; Wolman and Page 2002; Wolman 2002; Dolowitz and Marsh 2000; Evans and Davies 1999; Stone 1999, 2000; Peters 1997), institutional transplantation (De Jong et al. 2002b), institutional transfer (Jacoby 2000) and policy diffusion (Knill 2005) are all linked to each other by referring to the same phenomenon, but have different connotations. Policy diffusion, for example, focuses on adoption patterns of a particular policy within and between countries but has little to say about how policies are altered during the process of diffusion (Stone 2004).

In the context of practitioner projects, it is helpful to distinguish between the transfer of knowledge and the transfer of experience (see section 4.2.1). Much of the knowledge that is of interest to practitioners is not necessarily codified and reflected, but often is rather 'raw' or anecdotal experience from activities. This poses particular challenges to knowledge transfer. Humpl (2004) differentiates between the two by emphasising the context- and person-dependency of experiences. He identifies the challenge of experience transfer as the transfer of person-bound knowledge and the de-contextualisation of practical insight and the challenge of knowledge transfer as the contextualisation of knowledge by the application of theoretical insights.

Most of the policy transfer literature concentrates on transfer between states in bilateral learning settings and on the receiving side of information. Most studies argue either implicitly or explicitly that the 'borrowers' and 'lenders' of knowledge do not usually change roles in uni-lateral settings (for example Rose 1993). Moreover, many look at transfer between 'more developed' countries to 'less developed' countries, which – from a global perspective – leads to a domination of practices from certain countries (such as Anglo-America or 'the West') (Roy 2011). In transnational cooperation, these relationships can be complex and turn transfer into a multidirectional and networked matter when actors classified as lenders also draw lessons, while regions classified as borrowers act as models for other systems (Dolowitz and Marsh 2000). Even more complex, in many transnational transfer situations, one cannot clearly identify borrowers and lenders, but partner countries gain and give away knowledge and experience at the same time. An exception may be programmes involving Eastern European countries, where a certain dominance of the West-East transfer may be found.

Policy transfer is usually treated as either the dependent or independent variable: processes of policy transfer can be explained or policy transfer is used to explain policy outcomes, but can also be treated as both the dependent and independent variable in a more comprehensive analysis (ibid.). Four dimensions of analysis dominate existing research: the comprehension of knowledge as well as the usefulness, speed and economy of transfer (Pérez-Nordtvedt et al. 2008). However, due to a lack of research into the dynamic relationship between the source, the recipient and the type of knowledge, analyses of the processes and determinants of knowledge transfer and thus of transferability remain fairly limited. The few studies that include process aspects (Salk and Simonin 2008; Lane et al. 2001; Simonin 1999a, 1999b; Inkpen and Dinur 1998; Szulanksi 1996; Inkpen and Crossan 1995; Hamel 1991) mostly focus only on single aspects and thus on fractional explanations for learning processes. Little is known about how practices, concepts and ideas are diffused, such as the fact that knowledge and best practice transfer is not a one-way activity, but a process of knowledge sharing, re-interpretation and re-creation that involves feedback loops (Sahlin-Andersson 1996; Szulanski 1996). Even fewer studies make an effort to put these processes into broader conceptual frameworks. Here again, the 'black box' phenomenon is encountered, and the ability of policy transfer literature to contribute to the understanding of transnational project processes is limited due to the lack of theory building. One of the few examples of a study that takes process aspects into account is the analysis of transfer processes between local governments by Wolman and Page (2002).

The transnational and inter-organisational character of projects increases the complexity of transfer as cultures influence how people process, interpret and make use of knowledge. Thus, knowledge recipients will have to *assess* the relevance and validity of the information to see if they can generalise within their conditions. One of the most significant limitations of knowledge transfer, therefore, lies in the obstacle of making the transfer object work under very dissimilar circumstances. Despite the common challenges in Europe, there are profoundly ingrained differences between countries in terms of administrative and political structures and cultures. In this respect, the term 'transplantation' is quite useful as the body analogy points to risks of the transplantation being rejected in the host body (De Jong and Mamadouh 2002). In some cases, there may be a transfer of policy knowledge, but not necessarily one of policy practice, when cooperation partners 'adopt lessons for symbolic purposes or as a strategic device to secure political support rather than as a result of improved understanding' (Stone 2004: 549).

Particularly in settings with diverging framework conditions under which the transfer object has come into practice and under which it is supposed to be implemented such as often encountered in transnational cooperation, it is helpful to conceptualise transfer as a process of de-contextualisation. Knowledge is often and experience is always contextualised and the more systemic and strategic knowledge is, the more difficult it is for partners to accept and work with 'foreign knowledge' in their contexts due to its often very tacit nature (Child 2001). Making it usable for others requires the abstraction from underlying principles, followed by the experimental application in the receiving partner's environment. Otherwise, one-way transfer runs the risk of being limited to single-loop learning when partners do not question their underlying assumptions. These processes of de- and re-contextualisation take much time and effort and thus present a powerful barrier to transnational cooperation.

CATEGORISING AND CLASSIFYING TRANSFER OBJECTS

A considerable share of policy transfer literature focuses on the categorisation and classification of transfer objects, which differ in terms of their transferability. One example is the categorisation of transfer objects into institutions, policies, procedures, ideas, attitudes, ideologies, justifications, and negative lessons by Dolowitz and March (2000).

De Jong and Mamadouh (2002) cluster transfer objects into three *levels of action*, which are helpful when judging the profoundness of transfer processes:

1   the *constitutional level* defines the whole set of legal and socio-cultural conditions, rules, norms and values that provide context in which decision-making processes and relations take place;
2   the *level of policy areas* includes the systems of legal, financial, political and organisational relations between various government units;

3   the *operational level* consists of the whole set of exploratory activities, procedures, techniques and administrative forms used by individuals within
constitutional and institutional frameworks and thus refers to concrete decision-
making.

Dolowitz and Marsh (2000, based on Rose 1993) identify four different *degrees
of transfer*, which help to conceptualise how ideas and concepts travel between
'borrowers' and 'lenders': copying, which implies direct and complete transfer;
emulation, which involves transfer of ideas behind a policy or a programme, with
adjustments to differing circumstances; combination, which represents mixtures
of several different policies; and inspiration, where a policy in another jurisdiction
may inspire a policy change, but where the final outcome does not actually draw
upon the original. Direct copying may be possible in cases where the transfer
object is, for example, a technology; in most cases of transnational cooperation it
is less likely due to the high context-dependency of transfer objects and the diversity of framework conditions (De Jong 2004; Stead et al. 2008).

The transfer of policy practice is often practised through the use of so-called 'best
practice' examples, which are also particularly popular in the INTERREG context
(see section 6.3). Another example is the differentiation of different 'components'
that can be linked to different degrees of transferability by the OECD (2001).

- Components of *medium visibility* such as technologies, operating rules and
  methods have a high potential to be transferred.
- Components of *low visibility* – ideas, principles of action and philosophies –
  are harder to transfer. They are often too complex to be of relevance for
  actors in other settings.
- Similarly, components of *high visibility*, such as programmes, modes of organisation and institutions again tend to be very specific to local contexts and are
  thus also challenging to transfer to other contexts.

LIMITATIONS TO TRANSFERABILITY

As any transfer depends on the actors involved, on the acceptance of the transfer
object by the host country and its policy outcome, these aspects influence its success. Thus, it is practically impossible to ultimately determine general success
factors for policy transfer, not to speak of the difficulties of assessing 'success' in
general. As indicated above, conceptualisations of how processes of policy transfer take place are more than limited, but, similar to organisational studies, some
favourable conditions have been identified.

According to Dolowitz (2009), the type of transfer and the degree of information obtained depends on who is involved in the process and where in the
policy-making process a transfer occurs. Emulation may be more crucial at the
agenda-setting stage while copying or combining are more applicable at the policy formulation or implementation stage. Politicians, for example, may tend to
look for 'quick-fix' solutions and thus rely on copying and emulation, as they are

less likely to be interested in the detail of a policy or programmes than in learning about ideas, rhetoric and strategies. Civil servants, on the other hand, may be more interested in mixtures, more likely to engage in deeper forms of learning due to their time horizons and their task-focus (Dolowitz and Marsh 2000).

Dolowitz and Marsh (2000) make out three factors of failure:

1   insufficient information about policy/instrument and how it operates in the country from which it is being transferred (*uninformed transfer*);
2   crucial elements of what made policy or structure a success in originating country have not been transferred (*incomplete transfer*);
3   insufficient attention paid to differences between economic, social, political and ideological contexts in transferring and borrowing country (*inappropriate transfer*).

De Jong et al. (2002) show that cases in which the transfer tried to copy the full version of a model from elsewhere proved to be much more complicated than cases in which the transfer was more adaptive. They also show that loosely defined models or multiple models are easier to transfer than one definite model.

Although in many cases knowledge transfer is a form of learning (Wolman and Page 2002), not all transfer needs to include learning. Drawing on international relations literature, Stone (2004: 546) argues that transfer is likely to be more effective where learning has also taken place: 'learning can make the difference between successful transfer as opposed to inappropriate, uninformed or incomplete transfer'. She argues that learning processes should precede transfer processes: 'When consensual knowledge is developed at a transnational level, the potential exists for the exchange of ideas providing impetus for policy transfer' (ibid.: 548).

Another interesting aspect is the effort of transfer. In this respect, the 'stickiness of knowledge' refers to how much effort needs to be put into the process of knowledge transfer (Szulanski 1996). The more effort is required, the more 'sticky' the transfer becomes. The idea that most knowledge transfer is more or less sticky is becoming more and more accepted and many studies report on empirical evidence that suggests that knowledge transfers do not run by themselves, but that success records are rather low.

'Knowledge stickiness' can occur during different cooperation phases (Szulanski 1996):

- *Initiation stickiness* is the difficulty in recognising opportunities to transfer and in acting upon them.
- *Implementation stickiness* can be found after the decision has been taken to transfer knowledge and when the attention shifts to the exchange of information and resources between source and recipient. The stickiness depends on how challenging the communication gap between source and recipient is. This involves communication problems such as language, coding schemes and cultural differences.
- *Ramp-up stickiness* can play a role when the recipient begins to use the acquired knowledge. The stickiness depends on the number and seriousness of

the unexpected problems and the effort required to resolve them. The difficulty is again likely to depend on the degree of causal ambiguity of the practice.

- *Integration stickiness*, finally, can lead to the abandonment of the practice if it cannot easily blend in with the organisational reality. The stickiness then depends on the effort required to remove obstacles and to deal with challenges to the routinisation of the new practice.

Causes of the stickiness of knowledge are the nature of knowledge to be transferred, the characteristics of the knowledge recipient and the nature of the relational context. Among the characteristics of the knowledge, causal ambiguity and a lack of know-how and know-why are most influential (Szulanski and Capetta 2008). Moreover, unproven knowledge can be challenging, that is knowledge that has only been proven on a limited scale or scope. Among the characteristics of the recipient and the relational context between parties, the following aspects were identified as influential (ibid.):

- source lacks motivation: for fear of losing ownership or superiority;
- source not perceived as trustworthy;
- recipient lacks motivation: 'not invented here-syndrome';
- recipient lacks absorptive capacity: mainly a function related to prior knowledge. Recipient with a lack would be less likely to recognise value of new knowledge, to recreate and apply it successfully;
- recipient lacks retentive capacity: ability to institutionalise utilisation of new knowledge;
- arduous relationship between the source and the recipient: strength of the tie;
- barren context: organisational context should be fertile to transfer.

Szulanski and Capetta (2008: 528) note that more research is needed to unravel the micro-mechanisms of knowledge transfer and advocate to open the *black box of knowledge transfer* that 'deserves, more than many other organizational phenomena, to be opened and scrutinized to understand mediating processes'.

B Exchanging experience from pilot projects

The parallel implementation of individual pilot projects allows the direct exchange of newly made experiences. Project partners regularly update each other on relevant developments, advancements and insights during project meetings. As most projects make use of rotating meeting locations, they combine meetings with 'study trips' to the local pilot project. This allows transnational partners to experience the variety of projects in a more direct way than by narration only and to benefit from more cases than their own.

Thus, despite being relevant for the local context of partners, pilot projects can be interesting to cooperation partners engaged in related activities. In some projects, concepts are implemented at different speeds or to different degrees at different locations and offer the potential to observe and learn from faster and more innovative partners. However, if people are exposed to only a limited selection of actions and experiences, they tend to build up contorted perceptions and to over-generalise results (Verres 1979). Hence, also the individual's previous knowledge and understanding determine what aspects of experience are perceived.

The insight that implementation is possible by way of different approaches helps partners to think in alternatives and to become aware of their own deeply ingrained assumptions. Realising different framework conditions and their effects is necessary to understand the context and 'inner logic' of other pilot projects. This helps to avoid 'inappropriate transfer' (Stone 2004) when insufficient attention is paid to differences in economic, social, political and ideological contexts. It assists in assessing the chances of knowledge transfer and in concluding how conditions and instruments at home need to be adapted. Being aware of varying background situations also increases projects' potential function as 'good practice' examples (Sahlin-Andersson 1996).

In this context, the concept of '*observational learning*' is particularly interesting. In his 'Social Learning Theory', Bandura (1979) stresses that learning impulses can originate from both direct and indirect experience. The latter he calls 'observational learning' (or 'abstract modelling'). Social-cognitive learning theory distinguishes between two processes of disseminating social innovation: (1) getting to know new repertoires of behaviour and (2) their actual implementation (Verres 1979). Observational learning is of particular relevance for the first: people can acquire extensive behavioural patterns through observation of others without having to build them in a tedious trial-and-error process. Observing can abbreviate processes of acquisition and helps in disseminating new ideas and practices. It thus plays a significant role in supporting innovation: it can stimulate creativity by exposing observers to a variety of models that causes them to adopt combinations of characteristics or styles and it can be stimulated more directly by modelling unconventional responses to common situations (Hergenhahn and Olson 2005). Models exemplify and legitimise innovative practice, but they also accelerate them by encouraging the adaptation of behaviour. Observers hesitant to adopt new behaviour might lose their inhibitions by observing how others gain advantages.

Social-cognitive theory identifies a range of stimuli that activate observational learning and its implementation (Bandura 1979), which are comparable to the concept of 'absorptive capacity' in organisational studies:

1 *attentional processes* determine what is observed (dependent on maturation and previous experience);
2 *retention processes* are determined by verbal abilities and the ability to store and utilise experiences in the future;
3 *behavioural production processes*, including the ability to reproduce a skill, translate that which has been learned from observation into behaviour;
4 *motivational processes*, as learning occurs continuously, determine which aspects of learned responses are translated into action.

The specific set-up of pilot project portfolios impacts on the 'intensity of cooperation' and 'observational learning', as projects oriented towards a general exchange of experiences or those that are simply an umbrella for a series of local subprojects are less conducive to learning than projects in which partners jointly develop or implement a solution or strategy (Böhme et al. 2003; Colomb 2007).

4.3.3 Phase 3: Reflecting on New Experience and Knowledge

Reflection is fundamental to learning, as it is not possible to learn from actions without being aware of the consequences of one's own behaviour (Keen et al. 2005; Raelin 2001). In order to make meaning of experience and to translate them into more or less abstract concepts that will serve as guides for new situations, learners are required to reflect upon their experiences. An environment that supports reflection and reflexivity is an influential basis for learning.

Reflection can be defined as 'the practice of periodically stepping back to ponder the meaning to self and others in one's immediate environment about what has recently transpired' (ibid.: 11). When different ideas are reviewed for solving a problem, content reflection takes place. When procedures and assumptions are questioned and thus the way problems are solved, process reflection happens, while premise reflection refers to the questioning of presuppositions (Mezirow 1991). Reflection can also be triggered by the feedback from others when people become receptive to alternative ways of reasoning and behaving (Raelin 2001). Feedback from peers helps in detecting potential 'judgement errors' that are caused, for example, by untested assumptions, biases and 'normal cognitive processing', that is when people look for similarities instead of differences between past experiences and new challenges (ibid.). Research in the field of international projects showed that learning processes were more successful when project partners had the opportunity to jointly reflect on practical experiences (Vinke-de Kruijf et al. 2013).

Reflective practice thus increases the quality of learning in projects, but acquiring habits of reflective practice in projects can be highly challenging: 'in the typical, task-centred project where short-term pressures prevail, it is not an easy task to shift the focus from action to reflection' (Ayas and Zenuk 2001: 74). Learning tools such as dialogue, storytelling (see for example Cziarnawska 1997) and exercises for team building, but also informal contacts, study visits and small-group work serve as practices for reflecting on task and team related aspects (Keen et al. 2005). Useful skills include the ability to reflect in general and to expose one's reasoning and encourage others to be inquisitive. Also systemic and collective reflection and recording the lessons learned, for example by evaluations and concluding reports, increase reflective practice and thereby project performance (Ayas and Zenuk 2001).

Another barrier to reflection is the limited number of observations in a project while its environment is complex and changing (see below).

4.3.4 *Phase 4: Abstraction*

Transnational learning takes place through a constructive integration of the individual inputs. Joint learning processes involve some form of 'reorientation', when partners go beyond what they know already, during which their fundamental beliefs and ideas behind policy approaches may change (Hall 1993). The outcome of these learning processes, particularly in cases of double-loop learning and ground-breaking innovations, cannot be known with certainty in advance because they involve 'a dynamic, non-linear, and inductive process of joint discovery that is contingent on behaviour, cognitive, and administrative factors, as well as luck' (Lubatkin et al. 2001: 1374). A single organisation's learning benefit is then difficult to isolate, as it will be embedded in the context of the cooperation (Knight 2002).

In transnational contexts, knowledge abstraction is thus a complex matter and faces many challenges. To make sure that knowledge gains and learning effects are retained and made available to the whole partnership, instead of being dispersed as soon as the project is dissolved, they need to be codified. This challenge is also called '*knowledge sedimentation*'; the preservation of knowledge for use after a temporary organisation ceases to exist (Grabher 2004b). However, in the reality of project life, knowledge codification, preservation and consolidation are undermined by the very character of projects as temporary organisations and their prevalent present-time character. These features are contrary to long-term planning and render the issue of knowledge codification less relevant for their members than for members of permanent organisations (Bakker and Janowicz-Panjaitan 2009; Mainemelis 2001).

Moreover, much of the knowledge generated in project activities is embedded in the tacit experiences of group members. To be of transnational relevance and transferability, projects need to raise their knowledge gains and lessons learned to a level of higher abstraction (Di Vicenzo and Mascia 2008). A particular challenge to knowledge abstraction in transnational cooperation is posed by the underlying problem of knowledge generation from a limited number of pilot projects. Knowledge production mainly takes place by 'distilling' transnational knowledge from the project's case studies that serve as its 'evidence base' (Potter 2004) and thus challenges projects with learning from samples of one or fewer (March et al. 1991), which complicates the identification of causality and the drawing of inferences (Prencipe and Tell 2001). As these case studies usually are of a low level of external validity, the identification of causality and the drawing of inferences are challenging (ibid.). This means that lessons cannot easily be transferred to other actors and settings, and, therefore, case-based projects pose inherent obstacles to transnational transferability and durability. Consequently, knowledge production of higher validity requires the generalisation of case-based knowledge relating to specific conditions and 'de-contextualisation' for the external use of findings (Potter 2004).

Difficulties of organisations to *benefit* from project-based learning have been highlighted by a variety of authors (cf. Scarborough et al. 2004; Grabher 2004b;

Sahlin-Andersson 2002; Keegan and Turner 2001). This is the paradox of project-based learning: while individual projects may offer an environment conducive to learning, they may also create strong barriers to the continuity of learning beyond the project boundaries (Ayas and Zenuik 2001; Sydow et al. 2004). The subsequent implementation of project results by relevant organisations can then again face the systemic effect of 'learning boundaries' that are created when learning produces new shared practices at project level. These, in turn, reinforce the division between project practices and practices elsewhere in the organisation, thereby limiting the transfer of learning from the project to the rest of the organisation (Scarborough et al. 2004).

From their learning typologies and further empirical research, Prencipe and Tell (2001) developed a matrix of *learning mechanisms in projects* (see Table 4.3). The analysis of the horizontal dimension allows an assessment of the identified learning processes on which a project is based. Along the vertical dimension, the matrix maps the mechanisms on the different learning levels, of which the individual and group levels are shown in Table 4.3.

Based on empirical evidence, the authors were able to identify 'learning landscapes', of which the 'explorer type' or L-shaped landscaped is marked in

Table 4.3 Learning mechanisms in projects

Learning processes			
Level of analysis	Experience accumulation	Knowledge articulation	Knowledge codification
Individual	Practical work in local projects	Figurative thinking	Journal
	Job rotation	'Thinking aloud'	Reporting system
	Training	Scribbling notes	Individual systems design
Group/ project	Person-to-person communication	Brainstorming sessions	Meeting minutes
	Informal encounters	Formal project reviews	Milestones/deadlines
	Imitation	Debriefing meetings	Pilot project evaluation
		Lessons learned and/ or post-mortem meetings	Case writing
		Intra-project correspondence	Project report
			Guidebooks, handbooks, etc.
			Project plan/audit
			Intra-project lessons learned database
			Project history files

Source: adapted from Prencipe and Tell 2001

Table 4.3 (see darker area). Projects following this learning approach emphasise experience accumulation processes and knowledge transfer through people-to-people communication and lack formal learning mechanisms (phase 1 and 2). When projects aim at expressing (new) knowledge in various meeting formats, they make the step towards knowledge articulation ('navigator type', phase 3 and 4). Finally, projects including mechanisms to write down and retain knowledge move towards knowledge codification, which allows the access and exploitation for other members of the involved organisations and beyond ('exploiter type', phase 4).

An approach that actively embraces the inevitable diversity of project knowledge is that of 'cluster evaluation' (Sanders 1997). It was developed for programmes consisting of relatively autonomous projects, which follow the same general objective but implement very different measures. These can form clusters according to similar objectives, strategies or target groups. The approach allows the deduction of generalised insights for practitioners and decision-makers. The key methodological components of cluster evaluation include (Potter 2004; Sanders 1997):

1   Thinking in thematic rather than in geographical categories helps to formulate common findings and leads to overcoming the limitations of case-based experiences; groups of pilot projects can be analysed comparatively to find common links at project level. This helps to identify success and failure.
2   Communication and dialogue: group interaction and investigation are of major importance; knowledge and experience are exchanged (with recording and processing of relevant verbal information).
3   Facilitation: to keep the discussions going, to ensure equal participation and to steer the process of de-contextualisation using communication and visualisation techniques. Facilitators who can broker between diverse interests can support learning processes (Crona and Parker 2012).
4   Negotiation and validation: partners discuss and negotiate what can be learned from both individual and common experiences (by abstracting knowledge and deducting case-based peculiarities). Learning from the results of projects with similar strategies is enhanced by in-time interpretation and discussion of results with all project partners. Final results are fed back to partners to ensure specificity and completeness.
5   Participation: transferable findings are developed jointly and subsequently 'owned' by the project. This might raise the chances that results are actually used by partners.

A study that looked at learning from pilot projects in the energy sector found that most of the above-mentioned aspects also support 'convergent learning', that is, alignment of perspectives and actions as well as organisational adjustments and the direct application of pilot project results (van Mierlo 2012).

4.4 Conclusion

An analytical model was developed to systematically analyse knowledge development and learning processes in transnational projects. To this end, existing models of learning in alliances and temporary organisations were combined and amended. These include three different frameworks for researching inter-organisational cooperation (Salk and Simonin 2008; Dietrich et al. 2007; Lundin and Söderblom 1995). As these analytical models are concerned with *structural properties* of alliances and projects and how these influence learning outcomes, they inform the structural framework for transnational learning that lays the basis for project processes.

From the discussed existing models, the analytical model for transnational learning and knowledge development takes up the dimensions of partnership-specific, knowledge-specific and the two project strategy-specific variables of goals and action. These represent umbrellas of more specific variables, such as the institutional background of partners, homogeneity and heterogeneity in partnerships, knowledge types and characteristics, as well as project objectives, tasks and methods.

With the help of process models for learning from educational and organisational studies, a *process perspective* is integrated into the analysis of transnational learning and the 'black box' of project processes opened up. *Experiential learning models* add the dimension of 'experience', which is particularly relevant in the context of INTERREG projects, where practical experience plays an important role and projects practically test and demonstrate concepts, strategies and findings in 'pilot projects'. These models explain how 'raw experience' changes existing knowledge stocks and adds new knowledge. Still, they only focus on individual learning.

Models of knowledge creation developed in organisational studies enrich the understanding of transnational learning by adding both a process perspective and social aspects to knowledge development, concepts of relevant phases and learning mechanisms that support these processes. They provide ideas for how knowledge is developed over time. The SECI model includes a perspective on knowledge qualities and thereby – at least partly – embraces the idea of 'experiential knowledge'. It adds social aspects when conceptualising knowledge development through different stages of group interaction. The *model for inter-organisational knowledge creation* by Huelsmann et al. (2005) adds the differentiation of transfer and sharing, and existing and emerging knowledge. Finally, the identification of *organisational learning processes* and related learning mechanisms by Prencipe and Tell (2001) provide a framework for understanding the different steps of knowledge creation in more detail. The latter also allows mapping the learning approach of organisations with respect to their focus and potential outcomes.

All of these models provide valuable insights into how knowledge is created on the interface between individual and group processes. However, they all have their larger or smaller shortcomings and need to be seen in the context in which they were developed: mainly in *intra*-organisational settings that cannot necessarily be compared to *inter*-organisational and transnational settings. Particularly the fact that transnational projects are only short-term organisations within an inter-organisational setting challenges the 1:1 transferability of these models.

Still, the above-discussed process models for knowledge creation and learning allow the identification of six major steps of relevance for transnational cooperation projects. The analytical model is thus amended by a process perspective, which focuses on those phases that are directly related to project processes and the immediate project scope. *The process perspective of the analytical model thus consists of four major phases: (1) making experience, (2) exchanging experience, (3) reflecting, and (4) abstracting and conceptualising.* Projects do not necessarily run through these in strict chronological order and particularly the making and exchanging of experience can alternate.

In *phase 1*, the implementation of practical pilot projects allows the making of new experiences. Lessons learned can be used to improve the solution or to adjust management practices and policies. Pilot projects can be research-driven, management-driven or of political-entrepreneurial nature. Although they are meant to pursue the overall project purpose, they are often planned in advance and thus mainly based on the knowledge bases of project partners existing before the project start. Together with the often practised individual implementation of pilot projects, this limits their knowledge function to the production of individual and organisational experiential knowledge.

In *phase 2*, experience and knowledge are exchanged between project partners. This exchange can involve newly made experiences in pilot projects or existing knowledge stocks. The latter is related to knowledge transfer processes that involve 'knowledge senders' and 'knowledge receivers'. Knowledge transfer is a particularly complex matter in inter-cultural settings that involve different languages, values and framework conditions. Additionally, in the INTERREG context, transfer objects are often strongly context-dependent. These can be either re-used in similar contexts or raised to a level of higher generality through processes of de-contextualisation (see phase 4). The more systemic and strategic the knowledge to be transferred is, the more difficult these processes are.

A literature strand that is particularly informative with respect to processes of knowledge transfer in phase 2 is the field of 'policy transfer'. Although this is mainly concerned with transfer between governments in bilateral settings, the 'sending' and 'receiving' nature of participating entities is much stronger pronounced than in literature on organisational learning as is the international dimension. This literature strand provides categorisations of transfer objects and degrees of transfer that can be applied to other settings. Analysing differing degrees of transfer helps to understand how profound attempted exchange and transfer processes are and how ideas and concepts travel between those involved. Moreover, it identifies conditions and different degrees of knowledge transferability that influence the success of knowledge transfer as well as possible reasons for transfer failure. Szulanski's concept of 'stickiness' of knowledge adds a process perspective to the mostly static dimensions of other authors by providing explanations for transfer challenges in different project phases.

Additional to directly making new experiences in pilot projects, the parallel implementation of several pilot projects also allows the exchange of these 'raw' experiences. In many cases, similar or related concepts are implemented at different speeds or to different degrees at different locations so that project partners can learn from the different approaches. In this context, yet another theoretical concept is of help to conceptualise learning processes that relate to the experiences of others: 'observational learning' (Bandura 1979) assumes that learning can also originate from indirect experience.

In *phase 3*, project partners reflect on new experience and knowledge for interpretation and translation into more or less abstract concepts that serves as guides for new situations. Reflection is fundamental to learning in order to make sure that experiences are turned into new knowledge and that new knowledge is set in relation to existing knowledge stocks. Peers can trigger reflection when attention is called to alternative ways of reasoning and behaving, to untested assumptions, or biases. Although reflective practice increases the quality of learning in projects, it is extremely challenging in a task-centred and time-pressured project environment. Specific tools that enhance reflection and learning include systemic and collective reflection, dialogue, storytelling, small-group work, evaluations, concluding reports, team building, study visits, and informal contacts.

In *phase 4*, finally, project partners combine their individual knowledge gains and integrate these in the joint project result. This is when transnational lessons learned manifest in written products. It takes place by knowledge abstraction and generalisation to obtain knowledge of higher validity than case-based knowledge. It often involves processes of de-contextualisation, which are the more relevant, the more affected knowledge is context-dependent.

Complexity can be increased in international settings due to the high variety of identities and contexts in inter-organisational settings and thus of different knowledge bases. This and the institution of 'pilot projects' used in the INTERREG context increase the challenges of abstraction and de-contextualisation, which thus face two important constrictions. Firstly, the potential to de-contextualise knowledge set in specific socio-economic environments is limited as not all implicit knowledge embedded in the context can be translated into explicit, codifiable knowledge. Secondly, the limited number of pilot projects that function as the evidence base to transnational knowledge implies that knowledge generation is based on a low level of external validity. This limits the identification of causality, drawing of inferences and ultimately the transferability and durability of project results. In this context, the concept of 'cluster evaluation' provides a useful starting point for combining the experiences of several different, but related projects.

In order to make sure that knowledge gains and learning effects are retained and made available to the whole partnership and beyond, instead of being dispersed as soon as the project is dissolved, knowledge abstraction needs to be accompanied by knowledge codification. However, in the reality of project life, the very character of projects as temporary organisations and their prevalent present-time focus undermines knowledge codification, preservation and consolidation. These are contrary to long-term planning and render the issue of knowledge codification less relevant for their members than for members of permanent organisations.

5 From Theory to Reality
Applying the Model

This chapter presents the four case studies and discusses the processes of knowledge development and learning that took place in these selected cases. The case studies are analysed with the help of the analytical model for transnational knowledge development and learning developed in Chapter 4, starting with project structures and continuing with knowledge and learning processes. These findings are linked with a brief analysis of the projects' learning outcomes and a synopsis of the parameters with the largest effects on knowledge development and learning. This chapter concludes with the identification of a handful of parameters that proved to be effective across the case studies.

5.1 Description of the Analytical Approach and the Case Studies

With the purpose of the book being based on the identification of causal links as described in Chapter 1, a qualitative research strategy is applied that does justice to its rather complex and context-dependent nature. Four practical transnational projects form the starting point for a deeper understanding of the processes of knowledge transfer and learning by allowing direct observation of relevant events and interviews with relevant actors. The analysis suggests potential support factors, barriers and their causal relationships that are and can be relevant for transnational knowledge development and learning. Lessons learned about the projects are considered to be informative about the experience of the average project. The empirical approach is not concerned with collecting a 'representative sample' that allows for statistical extrapolations, but with generalising to theoretical propositions and thus allowing for analytical generalisations (Yin 2003).

Beyond the case studies, a mini-survey provides more comprehensive information on transnational working methods and ways to methodologically address the challenges of transnational cooperation.

Four case studies form the empirical basis of the study, of which three – at the time of the research – were finalised projects of the NWE INTERREG IIIB funding period, and one was an ongoing project funded during the NWE INTERREG IVB period. This approach allowed studying both aspects that emerge towards the final stages of a project or even after its termination (for example knowledge processing to produce final results) and dynamic and evolving aspects, which are better studied while taking place. The following section briefly presents the case studies with regard to their partnership, content, objectives and project strategy. To encourage interviewees to reflect as openly as possible on relevant cooperation processes, project names were anonymised.

Data collection methods included a document analysis of all available project documentation, interviews with a selected sample of project partners and participant observation in the ongoing project.

To ensure a certain variation of partner types, at least three partners from each project were interviewed, which included the Lead Partner where available. Interviews followed a guided conversation routine rather than a structured and rigid questionnaire. They were of an open-ended nature, in which respondents could reflect on facts as well as opinions and insights. Interviews had a particular strength with regard to the identification of causal relationships between the conditions of cooperation on the one hand and knowledge transfer and development on the other hand. This required personal knowledge from individuals directly involved in the process.

A *participant observation* was conducted in one ongoing transnational project, and four three-day project meetings were attended in various locations. This allowed observing the interaction of participants and their action, reasoning and communication to inform research findings. Additionally, one partner meeting and one larger project conference of two other projects were 'observed'.

Observations were recorded in a log that contained notes of all discussions and activities, including topics, questions, disagreements, challenges and their solutions. The log included both discussions that were led openly during meetings and more casual conversations during breaks and informal time.

WOOD

WOOD aimed at increasing biodiversity in forests while making integrated use of their economic potential. According to interviewees, the actual focus was on increasing the demand for (certified) wood products. Work had begun under another EU funding programme and continued from 2003 to 2007 under the INTERREG III B programme. The partnership was characterised by a mix of 13 national and regional bodies and NGOs from the Greater Region (border area between Luxembourg, Belgium, Germany and France), which represented a range of forest-related interests. By limiting the partnership to Greater Region, the project was able to deal with a contiguous forest region.

Fourteen different objectives were translated into three work packages:

1   The certification of wood according to an internationally recognised certification scheme, which had already been successfully implemented in France. Belgian partners had just started with its implementation while the scheme was unknown in Luxembourg. One of the French partners acted as work package leader to ensure that Belgian and Luxembourg partners implemented

lessons learned in the French context and thus to increase the overall quality of implementation. Benefits were expected not only for forest owners but also for the wood industry and primary production.

2   An increased quality of sustainable forestry with a focus on 'putting newest scientific insights into practice' with practical instruments and on knowledge development in the field of the influence of various forestry methods on the quality of soil, fauna and stands. Projected outputs were guidebooks and trainings for the different target groups.

3   PR measures for the certification scheme and wood products that resulted in a large variety of information material, excursions and a touristic route leading to forests and sites of interest for wood production.

A first application put forward to the NWE INTERREG IIIB programme had not been successful, but the partnership was encouraged to re-apply. After re-working its action plan and improving partners' assignment to project tasks, the project was finally accepted.

RIVERS

RIVERS was approved in the same call with WOOD, after having previously been rejected (due to too few tangible project results). The partnership of 13 partners from Germany, Luxembourg, France, the Netherlands and Belgium represented both NGOs and public authorities and ran from 2003 to 2008.

The project aimed at relating sustainable water management and the involvement of stakeholders and civil society as a means of enhancing political legitimacy. Project objectives included the creation of a knowledge base for the Rhine river basin, public information and consultation, the participation of different stakeholders, a general enhancement of cooperation in the river basin, new tools for public participation in water management, and the improvement of the ecological quality of the watercourse. This was translated into five 'components'.

The project had a German Lead Partner while a French NGO carried out project coordination. After around two-thirds of the project lifetime, the project coordinator left, and the Lead Partner took on responsibility for project management. According to project partners, the loss of the coordinator led to a subsequent loss of a substantial amount of knowledge and management expertise and the abandonment of several project elements (for example the evaluation of findings).

Although RIVERS could achieve some of its objectives at the transnational level, others were only achieved at the local level. The production of a publically accessible 'information and knowledge base' had to be postponed to an extension phase and was not created transnationally. No proof could be found that two of the project objectives, which were related to the rather general goals of enabling different stakeholders to become active players in water management and enhancing transnational cooperation in the river basin, could be achieved.

PARKS

This project brought together six partners from urban regions, including six planning authorities and one NGO. With a UK Lead Partner, the project included partners from the Netherlands, Luxembourg and Germany. PARKS succeeded a project from the INTERREG IIC period and was intended to test the expectations and conclusions developed by its predecessor through a series of measures and investments. The project applied in the very first call of the NWE INTERREG IIIB programme but was not approved due to the local character of pilot projects. PARKS then re-applied in the following call and introduced specific transnational methods. The project ran from 2003 to 2008.

PARKS set out to demonstrate the role of public spaces for sustainable development in densely populated areas. Moreover, the project meant to promote regional identity, to establish a 'planning through partnerships' approach and to develop transnational and regional learning processes.

PARKS had an elaborated methodic approach of transnational feedback sessions in which pilot projects were supposed to be jointly planned, designed and implemented and partners to act as 'advisor and consultants'. The project included an evaluation programme to monitor its influence on regional planning policy and on both individual and collective learning 'through innovative self-evaluation techniques'. Moreover, an external evaluator assessed individual and regional learning process. The project also included several thematic conferences to discuss sub-themes in more depth.

Due to the highly general nature of project objectives, it was not easy to get an impression of their achievement. The project aimed at jointly planning, designing and implementing a range of innovative pilot projects. Although these were implemented, project partners found that a joint approach to planning these had not been possible. The projected development of 'regional spatial plans' took place in two partner regions, but this was not promoted as a project result, and no lessons seem to have been concluded at transnational level. The objected good practice guidance did not go beyond a list of 'good practice' examples while the projected toolkit was not produced.

SEWAGE

SEWAGE set out to tackle the occurrence of pharmaceuticals in wastewater, which did not yet play a role in wastewater treatment. The project aimed at finding out whether a disconnected treatment of concentrated, pharmaceutically contaminated waste water could be a sustainable, cost-efficient method to reduce the discharge of these substances into aquatic systems. The planned results of the project were highly scientific, but also included practical solutions for treating different point sources and initiating a broad discussion on protection measures and responsibilities.

From 2008 to 2012, six partners formed the partnership from the Netherlands, Germany, Luxembourg, Switzerland, the UK and France. While two water boards installed and tested water treatment plants in the vicinity of hospitals, two research institutes conducted tests on temporary plants and two universities assisted with measurements. Actions were divided up into four areas: the analysis of the contribution of point sources to the overall mass of contamination, the planning and construction of pilot plants, the technical development and monitoring of the plants and the evaluation of joint results including their dissemination. This was translated into three work packages: (1) analysis, (2) testing technologies and (3) assessment.

The project followed an explorative design. While the first two work packages proceeded without further problems, the integrated assessment of the findings confronted the project with mixed results for different assessment perspectives. From a technical point of view, the project was able to recommend a decentralised approach to water treatment, but from an economic and overall environmental view, the assessment questioned the usefulness of tackling the overall problem by water treatment due to low reduction rates for the contaminated water.

Additional Project Cases Used in the Study

During the research project, it was possible to take part in individual project meetings of two additional INTERREG IVB projects. Although these were not full case studies, valuable insights from this experience were considered where appropriate.

The meetings included a regular partner meeting of a transnational project funded under the North Sea INTERREG IVB programme and a mid-term conference of another transnational project under the NWE INTERREG IVB programme. The first worked with maintenance aspects of open spaces while the latter was concerned with the improvement of public transport options.

5.2 People, Topics and Approaches: Project Structures

This section discusses the design of project structures along the structural parameters identified earlier as relevant for transnational cooperation and learning and, where possible, provides indications of their impact on learning processes. The four case studies serve as examples for the design of project properties and they illustrate possible links between these and the following processes. The discussion is guided by findings from the theory strands discussed in the previous chapter and presents the findings from the case studies at an amalgamated level.

Structural project properties are determined at the point of the project application and of more or less stable nature during the course of a project. In some cases, minor adjustments become necessary and may change the partnership or the applied strategy to achieve project results. Structural parameters are divided into three distinct groups: (1) those specific to the partners and the partnership, (2) those specific to the project topic and thus the relevant knowledge concerned and (3) those specific to the project strategy. Each of these groups consists of a variety of relevant parameters, which are discussed subsequently. The analysis of each of the parameters is again followed up by a summary of the major findings and a generalised overview that – by way of anticipating later findings – identifies inter-linkages of the relevant parameter both with other structural parameters and with process phases.

5.2.1 Partnerships

An actor-perspective on projects is a two-fold matter: partner structures represent a *static perspective* while partner behaviour represents a *dynamic perspective* on projects. The following analysis focuses on partner and partnership structures, which lay the basis for cooperation. Partnership-specific parameters are analysed in the following inter-related logic: (A) characteristics of the involved organisations, which include their institutional and sectoral background, (B) partner experience and knowledge and the motivation to cooperate including awareness of potential 'knowledge roles' and (C) composition of partnerships.

Research on knowledge management in organisations shows that the exchange, transfer and creation of knowledge are influenced by the ability and motivation of the relevant actors as well as on their opportunities to learn (Argote et al. 2003). Ability relates to previous experience and training while learning opportunities include options for observation and informal connections (ibid.). A certain degree of prior knowledge is necessary – at least among some of the partners – to be able to work on the project topic but is not sufficient for knowledge transfer and learning to take place. To develop new knowledge and learn, partners need to be aware of their learning potential and the knowledge pool held by the partnership. The awareness of potential 'knowledge roles' needs to be combined with an objective to learn or to teach to be translated into action. The composition of the partnership is particularly concerned with the balance of homogeneity and heterogeneity of partners that have different effects on cooperation.

A Partners' Institutional and Sectoral Background

Most actors in INTERREG B programmes are of public background, many representing local and regional authorities (BBSR 2009). Other actors include universities, research institutions, national authorities, non-governmental organisations (NGO) and some private partners. Similarly, partners in the four case studies were mainly public authorities, NGOs, universities and research institutions.

> There were different organisations, from NGOs over planning authorities to ministries. You find completely different working styles and tasks. This is also why we had different approaches to our topic. All in all, this was appreciated in the project, even if there had been some difficult moments. (Partner PARKS)

As inter-organisational learning literature focuses on homogeneous partnerships of private companies or governments, there are only a few clues on the potential impact of partners' institutional backgrounds on learning processes. Salk and Simonin (2008) argue that the key to understanding the capacity and motivation to learn is an organisations' governance form. According to their argument, centralised organisations do not learn as effectively in alliances as more decentralised organisations. Moreover, they point to the fact that the roles of individuals, groups and organisations differ with respect to knowledge recognition, acquisition and driving change in bureaucratically driven organisations. As public authorities are a majority of INTERREG partners, these projects are faced with specific preconditions to learning in partnership.

> For us as a ministry – we usually work in a rather abstract way – it was interesting to see how others work very practically at local level. Also the way they are organised and how they emphasise public participation is something we do not yet know in Luxembourg. (Partner PARKS)

As the case studies show, partners' institutional background also impacted on:

- work styles and routines: the degree to which meetings were structured, organised and 'bureaucratic' differed between NGOs and public authorities;
- communicative cultures (for example openness, dealing with conflict);
- partners' knowledge (for example university partners possessed more scientific knowledge, public partners and NGOs more practical experience);
- the capacity to implement project results (for example public authorities could install permanent infrastructure while research institutes and universities were more limited in implementing project results);
- partners' roles in the project (which are linked to their capacities, see previous point);
- partners' project objectives (which are related to their capacities and roles, see previous points).

The complexity of institutional types and their potential influence on project objectives, partner roles and their ability to implement project results was especially evident in the RIVER case study, where NGOs wanted to advance and expand public participation in water management, while public authorities, who had the capacity to implement these participatory approaches, preferred rather traditional and less cooperative participation styles and a less reformative approach in general. This impacted on the set-up and implementation of partners' pilot projects, which again had a direct bearing on knowledge processes (see 5.2.3).

Although many of the researchers in the SEWAGE project possessed strongly related knowledge and were highly experienced in laboratory settings, they found the development of practical recommendation from their project results highly challenging. This is a particular challenge for research-related projects. The other case studies had a higher representation of practitioners and thus, in theory, fewer barriers to the implementation of project results. Instead, the practical focus of these projects led to communication barriers due to a high share of tacit and context-dependent knowledge.

As one of the key characteristics of INTERREG projects is their *interdisciplinarity*, most projects include actors from different sectoral backgrounds. Interdisciplinarity was high in WOOD and RIVERS, involving different stakeholders along the relevant production chain, forestry and marketing experts (WOOD) and water management bodies, environmental organisations and participation activists (RIVERS). In contrast, *organisational diversity* was low in PARKS, where – except the Lead Partner (NGO) – all partners represented planning authorities. WOOD had the highest organisational diversity with NGOs cooperating with public authorities, an inter-professional association and an agency for regional economic development.

INTERREG projects are supposed to be of *multi-level* character, but this impacts on the capacity to implement results, as public authorities' competencies differ according to their spatial level (local, regional, national). Different spatial levels were represented in all projects, but diversity was the highest in RIVERS: organisations at European and international level cooperated with those working at national, federal state, regional and local level.

B Previous Experience and Motivation for Cooperation

From a resource perspective, a project's main resource is the expertise that its partners provide. This is the basis for exchanging knowledge and facilitates the process of new knowledge acquisition (Cohen and Levinthal 1990; Powell et al. 1996; Szulanski 1996) and lays the basis for 'observational learning' (Bandura 1979) that can help to make most use of the institution of pilot projects. Previous experience, skills, and motivational factors again determine the ability to perform 'observational learning'. The type and depth of expertise project partners depend on the organisational type and sector they represent as well as on their professional background and work experience. Obviously, project partners have different degrees of knowledge with respect to the project topic. Different knowledge types, including substantive knowledge (related to the problem and potential solutions), procedural knowledge (related to the organisation and management of the process) or political knowledge (related to the policy network) can be the cause for different knowledge bases (Schusler et al. 2003).

In the case studies, partnerships represented different levels of experience and knowledge, which impacted on the way partners defined, interpreted and implemented relevant concepts (*development stage*). These differences were partly linked to *national framework conditions* and traditions.

> In Luxembourg, for a long time we have not had actors who work conceptually in the planning field, except some private planning offices. We were simply interested in the topic. . . . It was all new to us. (Partner PARKS)

> It was new to us to cooperate with organisations specialised in water or even in water economy. (Partner RIVERS)

> Our objective was . . . to establish this certification scheme here in Luxembourg. We were behind the times. Germany, France and Belgium had all had the system for a while, but it did not exist here. (Partner WOOD)

In three of the case studies, individual partners were faced with topics they had no previous knowledge of, especially in respect to instruments and methods, but also to the degree of thematic specialisation. These *gradients* in partners' knowledge constituted knowledge boundaries and made communication more challenging but also provided transfer options and thus opportunities for learning in varied forms when partners found it valuable to access other partners' existing knowledge bases:

- *thematic transfer*, example: knowledge on sustainable forest management in WOOD;
- *transfer of instruments and methods*, example: French partners in WOOD transferred knowledge and experience on certification scheme;
- *transfer of values and attitudes*, example: NGO partners in RIVER attempted to transfer more inclusive approaches to public participation.

PARKS is an illustrative example of transnational knowledge transfer options. Cascade-like, the experiences of two German regions with 'regional parks' were used in the predecessor project, where a third German region initiated a process that aimed at introducing a regional park. This was followed with interest and 'observed' by partners from Luxembourg. During the PARKS project, the German region was able to build upon this process and developed the actual regional park while partners from Luxembourg used the German vision as a model for a similar process. However, interviews revealed that this knowledge transfer did not live up to its potential due to a lack of awareness of the 'knowledge roles' (see below).

Partners' knowledge bases pre-determine specific roles in the process of knowledge transfer and development, such as a 'knowledge sender' or an expert for a certain project task. However, how far partners are aware of their potential roles in knowledge exchange and development and thus willing to act accordingly is another question. On a variety of occasions, interviewees identified a certain learning intention and themselves as potential knowledge *receivers*, either because they lacked experience, or they attempted to advance existing experience. Knowledge gaps were identified with relation to more *specific thematic knowledge* (for example water economy) or *instruments* (for example PR or certification). Especially in the RIVERS project, partners lacked awareness of their potential roles of 'knowledge senders', while in the PARKS project, being aware of existing experience did not mean that partners were also aware of what they did not yet know.

In several of the case studies, the fact that some organisations and their staff possessed '*expert knowledge*', while others had more of a '*generalist background*' influenced the way the project exchanged and developed knowledge. In SEWAGE, specialised sectoral experts build on previous work in strongly related fields. These experts had clearly defined expectations and objectives, were target- and implementation-oriented from the outset and were ready to develop practical solutions. At the same time, the high share of experts from two strongly related disciplines also meant that the project was of a particular technical nature with an orientation towards nature sciences and of little interdisciplinarity.

In contrast, project partners in PARKS were generalists with experience from a broad variety of subjects. Involving generalists has its merits as they are often able to follow an integrated view, draw bridges to other disciplines and thus add benefit with their interdisciplinarity and openness to new aspects. Generalists can potentially bridge sectoral divides between experts when interdisciplinary approaches are required. However, in PARKS some of these generalists needed a significant amount of time to familiarise with the specific project topic to be target-oriented. Similarly, a study conducted by the interregional INTERREG programme shows that the high involvement of generalists in European projects rather than policy makers and experts prevents many projects from in-depth focussed and higher quality discussions on the project theme (INTERREG IVC 2013).

An interesting observation in two of the case studies was the different behaviour of experts and generalists towards identifying their potential 'knowledge roles'. Although the depth of generalists' knowledge was limited, in the PARKS project some of them had difficulties to clearly demarcate the scope of their knowledge and its limits. For the involved experts in SEWAGE, the situation was reversed: although their knowledge was deeper, they found it difficult to intellectually span to other areas. Still, they were more aware of the scope of their knowledge as well as its limitations.

Although a lack of awareness of one's knowledge role does not mean that knowledge is not transferred or developed, this is likely to be more coincidental. Interviews proved that partners tended to understand themselves in one of the two knowledge roles, but rarely saw the complexity and convertibility of their role. Still, 'senders' can potentially draw lessons on other aspects and 'receivers' can provide information in other areas. While one partner in PARKS, for example, was a front-runner of the 'regional park concept', they were not aware of their lack of knowledge on participatory processes. Another partner was highly experienced in the field of public participation, but new to regional governance aspects. In neither of the two cases did partners realise their receiving role.

As a variety of authors point out, the extrinsic and intrinsic motivation of project partners to cooperate and to learn has a bearing on their behaviour in the cooperation and learning process, while a lack of it can hinder the amount of knowledge transferred (Pérez-Nordtvedt et al. 2008; Osterloh and Frey 2000; Szulanski 1996). In transnational projects, the individual insight that cooperation holds more than just EU money is important. In their study on transnational INTERREG projects, Dühr and Nadin (2007) found that partners' motivation to cooperate was often based on their own local issues rather than transnational considerations. Although projects need to set common objectives in the application, the individual motivation to participate can vary. The case studies show that the fit of organisational objectives and the links between project objectives and organisational objectives influenced actors' identification with the project and how much they acted in concert.

> Let me start by saying this: for most partners the motivation to participate was to have access to funds for their local projects . . . It was the same for us. (Partner RIVERS)

A variety of interviewees identified their local benefit (*quotation above*) as a starting point for cooperation, which limited their identification with the overall project.

> You can also say that transnational cooperation makes your own project play in a different league. When the transnational feedback session took place in our region, motivating high-ranking public servants to give a presentation at the meeting was not a problem at all. (Partner PARKS)

Although PARKS was strongly built around allowing knowledge transfer and observational learning, interviewees did not identify transnational motivations for participation. Instead, local considerations and the hope that a European project would 'upgrade' local pilot projects among regional decision-makers and attract further funding to the region dominated partner motivations (*see quotation above*).

> Some partners really wanted to cooperate and would never start an action before getting feedback from the group. Others had to be reminded all the time that they needed to discuss with the others before getting started. (Lead Partner WOOD)

The motivation to participate in a transnational cooperation project can be based on local considerations, but it can also have a transnational dimension. In both WOOD and SEWAGE, at least some partners had more strategic objectives and explicitly wanted to achieve some 'change'. Transnational motivations included the *pooling of knowledge* and making use of *economies of scale and scope*, which highlights the resource-saving aspect of transnational cooperation. Although at least some partners in RIVERS were interested in options for '*observational learning*', accessing the different knowledge bases and speeding up regional development through the 'import of experience', this did not result in transnational knowledge processing.

> We partly had real economic savings by avoiding the bad experience and mistakes partners had made. Or the other way round: we learned how to do things better. (Partner WOOD)

WOOD and SEWAGE both contained exploitation and exploration aspects (Choo 1998); partners gained access to each other's expertise and joined forces, but also ventured into unknown areas together. Both projects also worked

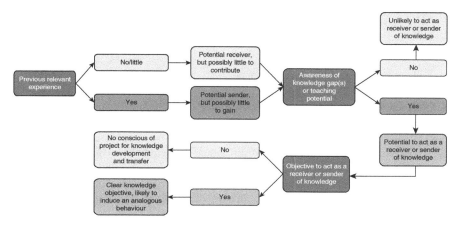

Figure 5.1 Influence of selected partner-specific parameters on knowledge transfer

Source: by author

towards *standardisation*. Partners in WOOD, in particular, were also interested in *accessing new markets*. In RIVERS and PARKS, strategies were – if at all – more oriented towards exploitation: 'outsiders with no vested interests are often better able to contribute fresh, innovative ideas to projects' (final activity report PARKS).

The motivation to participate had a direct impact on project results: the transnational dimension of the motivation in WOOD and SEWAGE allowed the creation of a common added value. On the contrary, while focusing much more on their regional benefit, partners in RIVERS and PARKS had difficulties in pointing out concrete knowledge creation processes and clear advantages from overall transnational cooperation.

Figure 5.1 summarises the relevance of previous experience and awareness of potential 'knowledge roles' for knowledge transfer and development. While the existence of previous experience related to the project topic turns partners into potential knowledge 'senders', its absence turns partners into potential knowledge 'receivers'. Transfer options then depend on the combination of partners.

The existence of potential 'senders' and 'receivers' is a necessary first step for knowledge transfer to happen but is in itself not sufficient without an analogous awareness of these roles. Without this, partners are unlikely to act upon their potential. If partners are aware of their knowledge potential but do not aim at contributing with expertise or at using learning options provided by transnational partners, knowledge transfer and development are unlikely to happen.

The motivation and objectives to participate in a project are influential on partners' behaviour. The case studies show – and this is supported by literature – that for project partners, their own local issues can be the main motivation for participating in a transnational project. Usually, project partners participate to make new experience, to advance their knowledge or to turn experience into practice.

On some occasions, however, their motivation may lie in the local benefit of local implementation.

C Composition of Partnerships

Social interaction is at the heart of transnational projects. Projects in the INTERREG Northwest Europe IVB cooperation area had an average of 10 partners per project (BBSR database[1]). The *composition of partners* is significant for the project course, its pilot projects as well as its potential for 'observation'. In a survey of learning in interregional projects, project partners identified inadequate partnerships as an important hampering factor to individual and collective learning processes (INTERREG IVC 2013). Policy transfer literature even states that the institutional fit is more decisive for transfer success than the degree of economic, political and socio-cultural resemblance between the involved partners (De Jong et al. 2002).

> The partnership was more or less coincidence. (Partner PARKS)

Compared to strategic alliances in business life, partnerships in transnational projects can be anything from a coincidental mixture of actors generally interested in the topic to a conscious choice of complementary actors of strategic fit to project objectives. A strategic partnership includes participants with diverse interests and different but complementary knowledge (for example substantial, strategic, political). The following section looks into the degree of similarity and differences in the four case studies and how this influenced knowledge development and learning.

Similarities between Partners

Inter-organisational learning theory suggests that similarities between partners affect alliance performance by facilitating the utilisation of tacit and articulated knowledge. These similarities can include a common reference frame, a general knowledge base, a unifying vision, strategic motivations, the ability to 'speak each others' language', organisational cultures and approaches to strategic decisions (Lubatkin et al. 2001; Jemison and Sitkin 1986). All of these ease cooperation. Several studies on transnational cooperation emphasise that also external factors, such as similar domestic frameworks are relevant for effective communication and that a partnership's geography influences cooperation intensity and the potential scope for learning (Colomb 2007; Dühr and Nadin 2007; Bachtler and Polverari 2007). Some go as far as suggesting spatial closeness between partners as a relevant support for achieving cooperation results (Lähteenmäki-Smith and Dubois 2006). Van Bueren et al. (2002) stress that problem-similarity supports cooperation. If projects include long-term cooperation approaches, similarities may become particularly relevant for joint agreements or standardisation efforts.

---

1   http://www.interreg.de (accessed on 20 May 2015).

We had close links with Luxembourg because we have so much in common. We also talk a common language, which is not the case with other German partners. (Partner from Saarland in PARKS)

This region is an entity from a forestry point of view. . . . That was a clear advantage for the project. The only difference was that some partners were more advanced in the process and others less. (Partner WOOD)

In the case studies, institutional, thematic or geographical similarities, as well as similarities in the knowledge base and similar interests and objectives created common reference frames and the ability to speak the same language:

- Similar professional and institutional backgrounds eased cooperation and transfer in SEWAGE and WOOD and partly in PARKS, but less so in RIVERS.
- Geographical similarities supported the existence of problem-similarity and thus the relevance of transnational cooperation in WOOD, but did not support transnational knowledge development in RIVERS.
- Similar knowledge bases with simultaneous differences in experiences and development stages in supported the exchange and development of knowledge in SEWAGE and WOOD.
- Similar interests and objectives gave SEWAGE the necessary structure and focus.

If at all, these similarities supported the overall cooperation process, but they never worked as a guarantee for a successful learning process.

The existence of spatial or functional linkages influences general inter-dependencies between partners, potential topics and objectives and their division of labour (for example along a river) and can thus affect results (Böhme et al. 2003). Lähteenmäki-Smith and Dubois (2006) argue that the transnationality of partner structures is mirrored in the transnationality of the output, or in other words that geographical networks are more likely to produce 'strong' cooperation results. This could not directly be confirmed by the case studies, where the trans-nationality of partner locations and potential inter-dependencies between them and their pilot projects was no guarantee for strong transnational outputs (see section 5.4). Still, similarities in transnational cooperation can only ever exist to a limited degree. The case studies also indicate that (a combination of) certain similarities may outbalance the lack of others.

According to the spatial characterisation of transnational projects by Böhme et al. (2003), both WOOD and RIVERS were geographical networks. Spatial proximity helped the 'regional cooperation' in WOODS to work suc-cessfully. RIVERS, however, an 'axial cooperation' along the river Rhine, was not able to balance out other obstructive factors that prevented the project from producing common results. PARKS was based on a few 'common geographical aspects' (shrinking and growing areas), and ensuring a degree of transnationality of outcomes and results was highly challenging. An improvement in the urban landscape in London, for example, did not require and was not necessarily linked to one in Frankfurt. SEWAGE was based on 'functional proximity' (same environmental challenges) and is an example for how little geographical factors can matter. It was strong on the process- and result-side, had a strong similarity of the problem and partners' professional backgrounds, made use of a genuine transnational strategy and methods, may have had advantages for knowledge processing due to highly comparable knowledge types and was able to outbalance a lack of geographically linked partner structures and locations.

Similarities and links in partners' knowledge bases may shape learning processes as well as their motivation and ability to make use of the knowledge obtained (Lane and Lubatkin 1998). An example for a tight relation between partners' knowledge bases was the SEWAGE project. Project partners drew upon their knowledge base to engage in knowledge development through project actions ('exploration'). 'Experts' shared a joint knowledge base and actively transferred experiences from pilot projects for joint analysis. In theory, each partner could have done the local analyses on their own; through cooperation, conclusions gained from cross-fertilisation. The access to each other's knowledge and the sum of the existing knowledge were thus only the basis for jointly developing entirely new knowledge.

Differences between Partners

While similarities lay the basis for communication and cooperation, to learn from and with each other, these need to be balanced by the aspect of *novelty*. 'Resource independence' of partners implies that they possess 'unique' knowledge so that partners benefit from accessing each other's knowledge (Lubatkin et al. 2001; Noteboom 2000; Child and Faulkner 1998). Differences in experience and backgrounds shape the knowledge diversity or 'cognitive heterogeneity' of a project (Di Vicenzo and Mascia 2008). They increase the 'reciprocal learning capacity' (Lubatkin et al. 2011) and the potential for knowledge integration by enriching the variety of options that are discussed (Scarborough et al. 2004). This can particularly support exploration strategies. In transnational cooperation, diversity is inherent and enhanced by the interdisciplinary and intercultural dimension of projects. As the evaluation of the INTERREG III programmes showed, different educational and professional backgrounds were the basis for sharing competence and skills and increased learning opportunities (Panteia et al. 2010). Different knowledge bases may be based on different thematic expertise, but also on different instrumental and practical experience. More experienced partners can move from knowledge transfer to development; they may learn *with* each rather than *from* each other.

> It is about making use of diversity to promote innovation in the old context. (Partner PARKS)

As outlined in section A and B, differences existed with respect to partners' institutional, sectoral, disciplinary character as well as in relation to spatial and knowledge levels. Diversity in transnational partnerships is a resource and added value for collaboration due to its potential for *innovation*: different partner perspectives and backgrounds can lead to greater innovation and creativity as well as a high degree of mutual learning (Iles and Hayers 1997). Innovative ideas and solutions can be tested that single partners would not be in the position of handling alone. Heterogeneous groups can turn it into an added value to exchange knowledge and use their capacity for creative problem solving if they manage to create a common platform, share their different sets of skills, knowledge and experiences and develop a joint language (Pérez-Nordtvedt et al. 2008; Reagans and Zuckerman 2001; Galison 1997). Di Vicenzo and Mascia were able to empirically prove a positive relationship between both cohesiveness and diversity in networks with project performance. The diversity perspective links similarity in knowledge with improved communication and commonality among project partners and heterogeneity with enhanced capacity for creative problem solving and the sharing of different sets of skills, information and experiences (Reagans and Zuckerman 2001). Heterogeneity is also referred to as 'network range' and can lead to a situation where projects 'have access to more knowledge components and will be able to mobilize and exploit different intellectual resources' (Di Vicenzo and Mascia 2008: 9f.).

> We should not gloss over the fact that the differences between partners and their framework conditions were a huge challenge and often led to annoyance. But the experience we gained is highly enriching. (Partner PARKS)

Still, diversity acts as a source for higher complexity, increased costs, decreased comprehension and speed as well as a potential barrier to transferability (Hachmann and Potter 2007; Huelsmann et al. 2005). If knowledge bases are significantly different, they can create knowledge boundaries that need to be overcome (Scarborough et al. 2004). Projects, therefore, need to manage their internal diversity.

Transnational projects are faced with a broad array of diversity: The complexity of interdisciplinary, cross-cultural, multilevel and inter-organisationally project groups potentially impact on cooperation. The disparities of member states in social and legal systems, governance, economic situations and administrative cultures potentially impact on the implementation of joint concepts and challenge the transferability of knowledge. Moreover, organisations of a similar type can work with different methodic approaches or have different values towards the subject matter. Different work frameworks can increase the diversity of internal and external interests related to the project (see section 5.2.1), which may make it difficult to find solutions that benefit all participants. As social development theory stresses, *cultural diversity* influences the perception of a challenge as well as its potential solutions. In the case studies, different national, and thus different cultural backgrounds added both value diversity (for example more democratic and open public participation approaches) and cognitive diversity, based on the different legal, historical, societal systems actors originate from (for example Napoleonic approach to water management in the Netherlands and France). Thus, the integrative and international component of transnational projects adds additional complexity to inter-organisational exchange and learning. Linguistic and cultural diversity can impede communication and thereby transnational learning processes.

Table 5.1 provides an overview on how variations and similarities of different structural variables took effect in the four case studies.

Table 5.1   Effects of similarities and differences on different parameters as found in the case studies

Parameter	Effect of similarities	Effect of differences	Effects in case studies (examples)
Institutional background	Supported ability 'to speak each other's language', supported transfer and result implementation due to similar competence; was more relevant in projects in later innovation phases	Enhanced creativity and questioning of usual routines	Similarities helped partners in SEWAGE to build comparable infrastructure and in PARKS to transfer knowledge, differences *could* have helped partners to advance their participation approaches in RIVERS
Sector, thematic orientation	Supported transfer and targeted working, supported implementation due to similar competence, supported ability 'to speak each other's language'	Enhanced creativity and questioning of usual routines	Similarities helped partners in SEWAGE to start quickly and work targeted, speak each other's language and in PARKS to transfer knowledge
Culture and framework conditions	Supported transfer and generalisation, supported ability 'to speak each other's language'	Enhanced creativity and questioning of usual routines	Differences created barriers to transfer in PARKS and problems to speak each other's language, but also inspired partners
Local challenges	Supported transfer, general focus and joint objectives, supported ability 'to speak each other's language'	Posed a challenge to general project direction, partner relations and common language	Similarities helped partners in SEWAGE to work focused and in WOOD to create a common reference frame; different regional challenges and a lack of joint problem description in PARKS left a lack of clarity on project objectives, a lack of

			similar challenges in RIVERS limited partner relations and common language
Expertise	Supported focused work and a quick start	Supported transfer	Similarities helped partners in SEWAGE to work focused and have a quick-start, but differences supported knowledge transfer and increased the quality of research; differences supported knowledge transfer in WOOD and PARKS
Partners' objectives	Supported a joint project objective and focus	Created only barriers in the long run	Similarities in SEWAGE permitted a common reference frame; differences limited exchange and cooperation in RIVERS
Pilot projects	Needed to be comparable for data collection purposes; needed a certain degree of similarity and fit the overall project for testing and demonstration purposes	Benefitted from a certain degree of difference, but needed complementarity for testing and demonstration purposes	Similarities helped partners in SEWAGE to assess pilots and come to joint conclusions; too large differences made joint conclusions in RIVERS and PARKS impossible

Source: by author

Interdependencies between Partners

Project partners not only included private forest owners, but also representatives from sawmills, architects and the whole production chain. Irrespective of INTERREG projects, there are different interest in the industry, not necessarily competition . . . but said very plainly, forest owners want to sell their wood, make as much money as possible, while sawmills want to spend as little as possible on it. (Partner WOOD)

In the WOOD project, partners were functionally interdependent. As Carlile (2004) points out, this can help in the development of joint objectives and results as the ability to share and assess knowledge depends on the actors' 'dependence' that is to achieve common goals, but can complicate cooperation when interests diverge. In both WOOD and RIVERS, certain organisations had opposing interests (see *quotation above*). Competition made cooperation more difficult as there was something at stake for actors. In RIVERS, NGOs and public administrations had different interests with relation to public participation, but NGOs were also dependent on public authorities. While the NGOs tried to increase and support participation in general, they needed the public authorities for the implementation of project results. Public administrations, on the other side, were rather reluctant to open up to participation processes and wanted to limit these to the legally compulsory level. In effect, while interdependencies were both helpful and hindering, different objectives and interests were detrimental to cooperation.

Project Partnerships: Conclusion

PARTNERS' BACKGROUND

Organisational characteristics of project partners vary in terms of their institutional and cultural background, their thematic and methodic focus and their degree of specialisation. INTERREG B projects usually represent a broad institutional and sectoral mix, which impacts on a variety of factors relevant to transnational learning, including work styles, previous knowledge and the capacity to implement project results. The considerable share of public partners in INTERREG projects may lead to particular challenges with respect to knowledge recognition and acquisition. Methods and mental frames may even differ within the same profession but between different countries. These differences create project diversity, which can both work as benefit and drawback.

Obviously, the optimal choice of transnational partners depends on project objectives, relevant tasks and planned results and thus has to be decided on a case-by-case basis. A project working towards creating policy recommendations, for example, requires partners who develop or deliver policies in their daily work, are in the position to influence policies or to transfer lessons learned to other actors. INTERREG programme structures, therefore, have a rather normative view on the organisational combination in a project. Project proposals are rejected if, for example, a partnership is 'too academic' and thus not able to implement project results, while other relevant partner characteristics such as their degree of specialisation or methodic approaches are not usually considered. As the case studies illustrate, partner choice was not necessarily based on strategic considerations and focused on potential knowledge gains, transferability and the implementation of results, but characterised by existing networks and coincidence.

The potential barrier that very practical, tacit and contextual knowledge poses is worth taking into account when dealing with funding programmes deliberately targeted at practitioners. The latter often work with a higher share of tacit knowledge and make less use of codified knowledge, whereas project partners with a stronger theoretical background face challenges when it comes to the implementation of project results.

Organisational characteristics potentially influence:[2]

- The composition of organisational characteristics particularly influences part-nership-specific parameters, but also knowledge-related (knowledge type and topics) and strategy-related parameters (objectives, methods).
- The kind of knowledge held, project objectives and the capacity to implement project results (see section 5.2.2).
- Certain organisational types are more likely to take over certain project roles and tasks, academic partners, for example, are usually better suited for analysis and advice, public authorities for the implementation of results.
- Cultural dispositions can influence working styles, communication styles, values, experience and context-related knowledge.
- With respect to the process, organisational characteristics can influence projects' communication, knowledge input and the likeliness of transfer through similarities in competencies and tasks (see section 5.3.2).
- Influences can also be expected with regard to implementing project results.

PREVIOUS EXPERIENCE AND MOTIVATION

The role of expertise for knowledge transfer and learning is recognised in literature and plays a role in INTERREG programme discussions related to project applica-tions. Relevant experiences can vary from 'experts' to 'novices' and from 'expert networks' to 'generalist networks' depending on the degree of specialisation both of the involved organisation and its member of staff. Previous experience influ-ences partners' ability for 'observational learning' and lays the basis for knowledge exchange and development. The use of this potential is, however, determined by a certain degree of learning intention and how strongly partners identify with their potential 'knowledge roles', how intensively they exchange and thus contribute to knowledge development. Communication then has a 'target' and thereby ultimately influences the willingness and motivation for knowledge processing. However, at the level of INTERREG programmes, there is no direct invitation to reflect part-ners' different knowledge roles and thus their contribution to the overall project.

The case studies suggest that intense but narrow knowledge and broader, more integrative knowledge both have their merits. Generalists and experts play dif-ferent roles in knowledge development and learning and complement each other. Experts, who look back on years of experience in a specialised topic and possess not only the know-what, but also the know-how and know-about (Lubatkin et al. 2001), are more likely to work in a target-oriented way right away and know what they can achieve and have a deeper knowledge base to share, but they may also be more limited and less flexible in reacting to challenges in related subject areas

---

2 This box summarises the findings of potential mutual influences between project parameters and phases and anticipates the analysis as a whole. Potential links are discussed in more detail as part of the discussion of the relevant parameter they relate to.

and in general more challenged in responding to the programmess requirement for integrative approaches. Generalists, on the other hand, may need more time to familiarise with the topic and may be less able to lead in-depth discussions, but may also be more open and flexible to integrate new aspects, bridge disciplines and provide a more integrative view on the subject.

As the case studies illustrate, the awareness of existing knowledge in one area does not necessarily include the awareness of non-existent knowledge in other areas. In interviews, project partners tended to understand themselves in one of the two roles, but rarely saw the complexity and convertibility of their role. Their awareness of potential 'knowledge roles' influenced the intensity of exchange and transfer processes, the division of labour and exchange methods in the case studies.

Previous knowledge and experience and the motivation to cooperate potentially influence:

- knowledge-related parameters, such as knowledge characteristics and the development stage of the relevant product or process (see section 5.2.2);
- strategy-related parameters, such as appropriate objectives and tasks, division of labour and partner integration (see section 5.2.3);
- the ability to design a pilot project in a way that it takes knowledge gaps as a starting point and not only implements existing knowledge;
- process aspects, such as the intensity of exchange process, cooperation and transfer, options for observational learning, feedback processes and ultimately the interest in knowledge processing (see section 5.3).

FINDING THE RIGHT BALANCE OF PARTNERS

Partners in transnational projects need to reconcile a certain degree of similarity and dissimilarity. Transnational projects build on the idea that cooperation benefits from the diversity of partners, their knowledge and experience. Aspects of novelty are required to make cooperation both attractive and beneficial and allow knowledge transfer. At the same time, they add complexity, increase costs and decrease cooperation speed and transferability. Interdependencies between partners can both support and impede cooperation. If not used constructively, diversity can limit the benefit of transnational exchange processes. Knowledge transfer, the development of joint knowledge, and joint actions can then turn into extremely challenging tasks. Thus, diversity in a project can act as a source of inspiration, innovation and complexity. It needs to be managed to allow projects to derive as much benefit from it as possible. In the INTERREG context, the barriers diversity can pose are not much discussed at programme or project level, which is also reflected in a lack of relevant methical advice to projects.

The case studies illustrate that some differences between partners served as a source for transfer of experiences and knowledge and enhanced the projects' innovative potential, while other differences led to challenges in project implementation and knowledge transfer. Knowledge transfer seemed to clearly benefit from similarities in background, but also from the difference in knowledge and expertise. Similar general knowledge bases, but different concrete knowledge appeared to help the exchange. Cognitive diversity and dissimilar problems together with non-clarified dissimilar values and objectives posed a barrier to transfer and cooperation in RIVERS and PARKS; in PARKS, differing framework conditions posed a further challenge. Heterogeneity may be more relevant to highly innovative projects and less so in projects focused on implementation. Although a gradient in partners' experience increases the chances for knowledge transfer and development, it is not a guarantee, as options for exchange and transfer are not always made use of. Neither does a lack of a knowledge gradient exclude knowledge development.

> It is about finding the right balance. Even if you only stay in touch with few of the partners later on and even if you cannot discuss at your level with all partners, diversity is important. But it also makes cooperating so much more difficult. (Partner PARKS)

When partners want to co-learn and create new knowledge, the 'right balance' between similarities and differences of partners has to be found (Lubatkin et al. 2001; Saxton 1997). Their 'strategic fit' determines the extent to which organisations can get along and realise anticipated synergies critical to a project's success (Saxton 1997). Partners' 'strategic fit' depends on the project topic and its main objectives and approaches (for example exploitation vs. exploration, Choo 1998) as well as its impact on other structural aspects and the overall knowledge transfer and development process. Therefore, a generalised success factor with respect to similarities and differences does not exist and the best fit has to be found on an individual basis. For those differences that may become barriers to transnational learning, ways may need to be found to overcome these obstacles and manage differences.

The functional partnership is potentially influenced by:

- partner-specific parameters, such as organisational differences among project partners, their cultural and professional background, different levels of expertise, motivations and objectives; it is thus strongly connected to the 'strategic fit' of partners and pilot projects;
- strategy-specific parameters: there are links to the general project orientation, as diversity supports innovation and creativity in an exploratory strategy, while similarities help to put results into practice and thus support exploitation or more implementation-oriented strategies (see section 5.2.3);
- different project tasks require different partner roles and backgrounds and their integration in the project at different starting points;

(continued)

(continued)

Functional relationships between partners potentially influence:

- partner-specific parameters such as partners' motivation to cooperate, and objectives;
- diversity can impact on some knowledge characteristics, especially knowledge complexity and the 'fit' of contextual knowledge;
- strategy-related parameters: the transnationality of the project, possibly the strategic fit of pilot projects, higher levels of diversity gain from clarity of objectives, tasks, and methods (see section 5.2.3);
- all relevant elements of knowledge transfer and processing (see section 5.33).

5.2.2 Knowledge

Knowledge-specific parameters are concerned with qualitative aspects of knowledge and strongly influenced by the project topic. Different knowledge types and characterisations affect the relevance of knowledge for transnational partners but also its transferability. Knowledge-specific parameters are analysed according to the following logic: (A) knowledge types and characteristics occurring in the case studies, and (B) a distinction between project types related to their knowledge-orientation and how these impacted on knowledge development and learning.

A Knowledge Types and Characteristics

The literature on inter-organisational learning emphasises the influence of knowledge and its features on the depth, pace and meaningfulness of knowledge transfer and development. The degree of partners' expertise influences the production of more shallow (know-what) or deeper forms of knowledge, including causal relationships (know-how, know-about, know-why) (Lubatkin et al. 2001). Moreover, the mix of partners and their knowledge impact on the effectiveness of the exchange, transfer and development of knowledge.

The further development of expert knowledge requires sufficient know-how; otherwise the learning process is limited to the transfer of existing knowledge. Retrospectively, it is relatively difficult to assess what knowledge partners possessed before the project start, but some indications with respect to knowledge depth could be found in the project documentation. Both WOOD and SEWAGE involved more than basic knowledge and partners were able to discuss causal relations quite profoundly. Project results reflect this. During the participant observation in SEWAGE, it was possible to detect a quite high degree of know-how, know-about and know-why among project partners, a high ability to exchange very informed views and to jointly develop new and very detailed knowledge about water treatment concepts. PARKS attempted to discuss causal relations on some occasions, but joint knowledge development was not a project objective as such and sufficient platforms for knowledge exchange and development were not provided. In RIVERS, very few causal links were made, and the project topic was little discussed as such.

> SEWAGE could work in all places; we do not need the water treatment plant to be located at [place name]. This project is about new research insights, which are relevant throughout Europe and not bound to certain locations. (Lead Partner SEWAGE)

Experience was related to *contextual* knowledge (for example planning systems, decision-making processes) or *context-free* knowledge (for example technical aspects). Particularly partners in PARKS felt they needed to understand the *different societal frameworks* to better understand each other's approaches and mindsets. When pilot projects were involved, experiential knowledge played a role, which often has a high degree of *tacitness*.

Complexity and specificity, but also validity, novelty and relatedness, uniqueness, value and actionability of knowledge are argued to impact on the pace, depth and meaningfulness of learning (Salk and Simonin 2008; Argote et al. 2003). Complexity in INTERREG projects is naturally high, as projects work in an interdisciplinary way and contribute to overriding questions, such as their relevance for regional development and the cooperation area. Complexity can increase causal ambiguity, which in turn can severely limit the comprehension and transferability of knowledge. Moreover, projects' context-dependency is often high, as they have a strong practical approach. Still, some of these characteristics can be influenced; clearly defined project objectives decrease complexity, a pilot project linked to very particular framework conditions increases knowledge specificity.

> I was quite confused by the structures in London and could not understand what they were trying to achieve, also because our cultural approaches are so different. (Partner PARKS)

> Many questions . . . are aiming at the long-term benefit of the cooperation: If we want to improve planning culture, we have to understand it first. . . . I think this might be called intercultural transferability: If we succeed in identifying real solutions for problems common to us all, we need to check which conditions apply. (Partner PARKS)

Projects dealt with different subjects and disciplines, ranging from forestry management over water management to open space in urban regions. Some focused on very practical issues and concrete instruments and methods, such as participatory methods or filtering techniques. It is not possible to systematically assess the knowledge characteristics of the four case studies, but some relevant hints could be identified:

- Practical knowledge was of high *actionability* and *novelty* to partners.
- A diverse pilot project portfolio decreased *relatedness* of relevant knowledge, which influenced how well partners could relate to each other and thus the fruitfulness of discussions.

- While the innovativeness of project topics and relevant methods led to the high *novelty* of the knowledge in PARKS, WOOD and SEWAGE, the *value* assigned to this knowledge affected partners' motivation to exchange and learn. In RIVERS, both the novelty and value of knowledge remained rather limited.
- In some cases, partners were faced with topics of a more constitutional type (for example transmissibility of rivers, regional identity to landscapes), which resulted in a high *complexity*, which influences knowledge transferability (Salk and Simonin 2008). High complexity coupled with a lack of relatedness of pilot projects made knowledge transfer and development difficult in RIVERS.
- *Specificity* was particularly high in RIVERS and PARKS due to a strong dependence on pilot projects on their framework conditions (for example cultural preconditions).
- The *validity* of the knowledge involved, but also its *relatedness* was very high in SEWAGE; all partners seemed familiar with relevant concepts and ways of thinking. As the other projects were less research-oriented, knowledge was often more of the know-what type, for example, design options or specific methods and instruments. Only in a few cases did these involve causal relationships.

Complexity was particularly high in PARKS, where partners discussed different philosophies and approaches and where the relatedness of specific knowledge and experience was notably low. Moreover, to increase the transferability of its highly context-dependent knowledge, the project let background knowledge dominate discussions, which again increased complexity. Together with a lack of specified objectives and tasks, this also led to low levels of actionability. As will be seen later, both PARKS and RIVERS faced difficulties when trying to produce their projected transnational results.

In terms of influencing knowledge characteristics, complexity can be reduced by narrowing the project subject with the help of concrete objectives and guiding questions. SEWAGE was characterised by high levels of concreteness and further reduced complexity by solely focusing on one source of substance emitter. Nevertheless, the development of a decision-making tool that involved factors beyond partners' expert knowledge overwhelmed the project later. The other three projects did not make attempts to reduce knowledge complexity.

In general, the character of knowledge in SEWAGE was different from the other projects; discussions were strongly following scientifically deduced arguments, with partners exchanging highly specialised knowledge. This led to a high degree of knowledge validity and thus transferability, which is less likely to achieve in cases where more social science-related knowledge or practical knowledge is involved.

With respect to the transnationality of the topic, the INTERREG programme for Northwest Europe has for a while distinguished between '*transnational issues*' and '*common issues*' (see section 2.5). Research has shown that in the case of 'common issues', learning is often limited to the regions involved and has less of a transnational dimension (Böhme et al. 2003; Lähteenmäki-Smith and Dubois 2006). This leads some authors to suggest that transnational projects worked better if they focused more on 'transnational issues' (Colomb 2007; Dühr and Nadin 2007). Again, the empirical analysis suggests a more differentiated examination. While WOOD and RIVERS were based on 'transnational issues', PARKS and SEWAGE were based on 'common issues' that could, in theory, have been solved without transnational cooperation. In RIVERS, the 'transnational issue' was even accompanied by a transnational partner structure. Despite these supportive preconditions, neither the project's processes nor its results turned out to be particularly transnational. It is likely that this was caused by a lack of transnationality of the project strategy. As will be seen later, there was no transnational division of labour, partners' commitment was largely confined to individual pilot projects, even 'common actions' did not involve all partners and pilot projects were of a weak vertical and horizontal fit. Consequently, transnational knowledge development was heavily limited (see section 5.3.6). Despite not pursuing a 'transnational issue', SEWAGE found a successful division of labour, equally involved all partners, strongly focused on its joint objective and common results and had pilot projects of a good vertical and horizontal fit.

B Innovation Phases in Cooperation Projects

In one or the other way, all INTERREG B projects deal with innovation, with developing and introducing products or processes that did not exist before. Classically, innovation can be understood as running through several distinct phases (for example Schoen et al. 2005). New ideas are generated and screened, translated into concrete concepts that are tested and assessed (for example by feasibility studies) and then implemented in real life. Similarly, policy-related innovation also runs through several phases, thereby influencing which type of transfer will be possible (Dolowitz 2009). For a transnational project, this may mean that the early phases of problem identification and agenda setting are subject to inspirational or knowledge emulation transfer processes while direct copying is more appropriate for implementation-oriented projects.

Innovation phases can also be linked to the concept of knowledge *exploration* and *exploitation* (Choo 1998). Explorative learning processes focus on discovery and experimentation to identify new goals and untapped opportunities and are of rather long-term nature. The pursuit of new knowledge can be more of a messy and chaotic process with no guarantee of success. Exploitation processes, on the other hand, describe learning through specialisation and building up of experience within the scope of existing goals and activities. Existing knowledge is productively used and short-term returns more certain (cf. March 1991). Brady and Davies (2004) see a connection between these two concepts and time aspects: organisations start with exploration activities and go over to exploitation activities. This could explain why in very innovative projects, the implementation aspect can be under-developed. The differentiation is useful for understanding the general orientation of knowledge creation in a project.

The fundamental difference between projects with a strong knowledge-orientation such as SEWAGE and those focusing on implementing practical outputs such as WOOD suggests that projects can be located along innovation phases. A focus on knowledge does not necessarily exclude implementation (in SEWAGE also infrastructure was built) but implies that large parts of the project focus on knowledge production. Similarly, an implementation-orientation does not exclude knowledge production. SEWAGE can be located in the early innovation stages (new ideas, screening and testing) and represents the explorative project type, but also included elements typical of later innovation stages (assessment, implementation). Cooperation was mainly focused on generating knowledge, and pilot projects served as relevant instruments. WOOD represents the exploitation approach and implementation-oriented project that included single knowledge-generation aspects.

> I continuously asked myself if this project was about joint insights, but I did not find an answer to this. I had the need to develop a joint result that could bring things forward. (Partner PARKS)

The cases of RIVERS and PARKS are more ambiguous. Both projects had a certain knowledge-orientation, which, however, lost focus during the process. At the same time, they had a strong focus on implementation through pilot projects. In RIVERS, interviewees were not able to make out concrete knowledge objectives. In PARKS, interviewees were not able to identify knowledge gains directly related to project objectives and knowledge gains at project level were not directly linked to pilot projects. Pilot projects were more or less limited to implementing existing knowledge at local and institutional level and had very few mutual ties, which limited the transferability of experience.

Knowledge: Conclusion

KNOWLEDGE TYPES AND CHARACTERISATIONS

Knowledge is an important resource, and its development is a central project process and an expression of a project's results. Beyond the content relation of knowledge, the choice of project topics, types and its characteristics can decisively influence the pace, depth and meaningfulness of knowledge transfer and development. Of particular relevance are complexity, specificity, validity, novelty, relatedness, value and actionability for learning processes in projects. Transnational INTERREG projects are mainly perceived in terms of dealing with programme priorities and less in terms of relevant knowledge types, which depend on the project topic and the type of partners involved. Many INTERREG B projects involve topics that are based on very different societal systems without partners necessarily being aware of related underlying assumptions and implicitness. The findings from the case studies support relevant literature, which identifies knowledge complexity as one of the most influential factors in the knowledge process.

It is itself influenced by the general diversity of transnational cooperation, but also by project strategies and the context-dependency of knowledge caused by the tacitness of knowledge and a high share of practitioners. The existence of different knowledge levels influenced the novelty and value of knowledge, whereas the relatedness of knowledge can be limited in highly diverse pilot project portfolios. Some of these characteristics may be managed, such as complexity that can be reduced with the help of concrete objectives.

Previous studies emphasise the role of 'transnational issues' as opposed to 'common issues', but this reduces the perspective to a limited range of eligible topics. In the case studies, the existence of transnational issues was not a guarantee of producing transnational project results and achieving transnational learning processes.

Knowledge characteristics can be influenced by:

- the partner mix and their knowledge and expertise (for example scientific knowledge vs. practical knowledge, the sectoral perspective of the organisations involved), diversity;
- project objectives and tasks: the more sophisticated a project strategy, the more know-how and deeper forms of knowledge are required, unspecific project objectives increase complexity (see section 5.2.3);
- the project topic.

Knowledge characteristics potentially influence:

- other structural parameters, such as motivation to exchange knowledge and learn (particularly the value, validity and novelty of knowledge);
- relatedness of knowledge affects how well partners relate to each other and their pilot projects in exchange and feedback processes (see section 5.3.2);
- tacitness and complexity influence transferability and ability to articulate experience.

INNOVATION PHASES IN TRANSNATIONAL PROJECTS

A perspective that has been largely missing in the INTERREG context is that of *innovation stages*. Although programmes include a programme priority on 'innovation', its conceptualisation has remained rather weak. As seen in the case studies, projects can involve more than one innovation phase and can theoretically span the whole process from idea generation over the development of concepts to their implementation and promotion.

A severe obstacle to innovation and particularly to transferable innovation, however, was found in those projects that strongly focused on individual pilot projects and disregarded joint work. The lack of connection between pilot projects and the overall project prevented the development of their exploitative and explorative aspects. It can be concluded that projects in different development stages and with different approaches have different requirements, run through different knowledge processes and thus need to be conceptualised differently. The strategic fit of the different project elements (for example pilot projects) to relevant innovation stages impacts on how much they can contribute to the project's cognitive interest.

The innovation-orientation of projects is potentially influenced by:

- partners' expertise.

The innovation-orientation of projects potentially influences:

- project strategy, such as objectives, tasks, methods, pilot projects' relation to the overall development stage (see section 5.2.3);
- type of learning process: creative idea generation and analyses allow learning about the subject while the deduction of concrete concepts and transnational coordination efforts focus learning on transnational governance. Very implementation-oriented projects concentrate on practical learning processes, which is possibly limited to single-loop learning.

5.2.3 Project Strategies

Parameters related to the project strategy deal with the direction, organisation and tasks of a project and provide the general backbone of its actions. The direct relationship between objectives and results determines the general project strategy. In the INTERREG context, project strategies have received limited attention. Strategy-related parameters are analysed according to the following logic: (A) to start with, a project requires a *vision* of what is to be accomplished and a set of *objectives* along which it can be assessed; (B) relevant *methods and tasks* help to achieve the objectives; (C) project partners need to find a *division of labour* that allows them to make the best use of their skills, and that links up single activities; (D) a very specific element of projects in the INTERREG context and many other transnational programmes is the institution of *pilot projects*. The way these are set up and do or do not serve project objectives influences the project's knowledge potential and impacts on the ease of transferability of project findings.

A Project Objectives

Analysing project objectives provides information on what is to be accomplished and how well a project has been planned in advance. The clarity of objectives has been identified as a key criterion for project success (Turner 2009; Nicolas and Steyn 2008; Slevin and Pinto 1987).

It was about doing things locally [and] integrating this into the joint overall objective. (Partner PARKS)

We never wanted to push our position by all means, but rather to jointly develop something that would advance all partners. Partners applying for such a project should be aware that they cannot pursue their own interest at the expense of others. (Partner WOOD)

With relation to project objectives, four specific aspects played a role in the case studies:

Two dimensions of project objectives: As transnational projects are multi-organisational, interdisciplinary and inter-cultural, the entirety of interests and intents for the project can be as varied as its members, and objectives have an *individual and a common dimension*. As the case studies highlight, it cannot be taken for granted that partners are aware of the benefit of transnational cooperation and that individual objectives are in line with overall project objectives or with each other. Although it is understandable that partners also pursue individual objectives and benefits, their motivation to learn or to teach is vital to the transnational learning process and their own 'reciprocal learning capacity' (Pérez-Nordtvedt et al. 2008; Lubatkin et al. 2001). 'Goal interdependence' (Lubatkin et al. 2001) means that participants realise that they can best achieve their goals by cooperation and that it is necessary for cooperative learning processes. In case of a lack of common objectives, it can only be estimated how much the resulting lack of identification with the project, lack of joint focus of activities and of motivation to relate to each other influences other parameters, but the case studies suggest that it can be considerable (see section 5.4).

The project aim was to build up a network. I do not know about more specific objectives, as I had not been a member of the founding team. What kind of objectives they had . . . you will probably find some nice formulations in the application. (Partner RIVERS)

The identification of common objectives requires project-internal discussions and the identification of a common ground that can lead to the development of joint results. Both WOODS and SEWAGE were to a large extent focused on common objectives. On the contrary, interviewees in RIVERS and PARKS had difficulties in defining a common overall objective (*see quotation above*). In such projects, very local and self-centred interests that do not acknowledge the potential of transnational exchange and working, partners' identification is more likely to remain with their particular pilot project and thus efforts towards processing common knowledge to produce common results are likely to be limited. For the Lead Partner in PARKS, the project dealt with participation aspects only while for other partners it also involved governance aspects and the development of 'regional parks'. The identification of potential 'knowledge roles' (see section 5.2.1) takes place within the thematic and methodic frame of a project. This means that options for transfer and joint knowledge development depend on certain congruence in understanding the project in terms of its content. A lack of this limited partners' ability to become aware of potential knowledge roles and thus hampered transfer and knowledge development.

Clarity of objectives: The clarity both of short- and long-term objectives guides the overall project process and thus influences learning effectiveness (Ayas and Zenuik 2001). A lack of focus led to inconsistencies in RIVERS and PARKS and negatively affected the logical progression through the project and thus knowledge transfer and development processes (see Table 5.2). Moreover, exchange processes were more difficult and diffuse.

Different meaning of the term 'objective': Although project objectives play a significant role, the conceptual difference in the meaning of the term has not been much clarified in the INTERREG context: it can refer to the tasks to be completed or to the overall achievements of the project, that is the envisaged change (Lundin and Söderblom 1995). Change or transition again can have two distinct meanings, both of which are relevant to the process and its outcome: (1) the actual transformation and distinctive change between 'before' and 'after' or (2) perceptions of the transformation or change among project participants and how this can be accomplished (ibid.). The latter is important to the inner functioning of projects as it focuses on perceptions of causal relationships and their consequences. The question, therefore, is how much the various partners identify with the general project vision and how well individual interests and objectives fit together to achieve 'goal interdependence' (Lubatkin et al. 2001). Table 5.2 summarises the quality of project objectives in the four case studies. Neither in RIVERS nor PARKS were interviewees able to identify the potential transition their project supported.

While SEWAGE's objective was very concrete, the other projects kept their objectives more open or even ambiguous. Both in WOOD and SEWAGE, sub-objectives clarified the transitional intent. In RIVERS, sub-objectives partly remained unclear, and none included transitional aspects. In PARKS, they involved only very fuzzy transitional aspects or were of extremely general character.

Clarity on the transition to be achieved requires awareness of *causal relationships* with respect to the topic. Indications could be found in WOOD and SEWAGE that a discussion on causal relationships had taken place. By testing and applying a variety of methods of very different relevance to the joint vision of sustainable forestry, WOOD was able to intensively and critically examine how to best achieve its objectives. Subsequently, several actions were amended or even abandoned. In SEWAGE, intense discussions about causal relationships took place, especially in relation to the conclusions from pilot testing. In the other two projects, transition played less of a role. In RIVERS, a critical reflection of the general division of labour and roles between NGOs and public authorities in participation processes would have supported the achievement of the project's objectives.

Table 5.2 Objectives in case studies

Clarity of objectives	Concreteness of common objectives	Sub-objectives	'Transition' aspects	Guiding question
WOOD	To develop joint and concrete actions to promote sustainable forestry → very open	15 sub-objectives clarified task; example: *'to introduce the certificate'*	Overall objective allows an idea of what was to be achieved	Problems and potential results addressed, quite heterogeneous, focus on implementation
RIVERS	To develop a sustainable, participatory and integrated water management → very open	Sub-objectives detailed but not necessarily clear; no information on instruments; example: *'to inform and consult the public [. . .] about the issues and challenges of sustainable water management'*	Overall objective allows an idea of what was to be achieved	Unclear, no problems and potential results addressed
PARKS	To recognise and demonstrate the vital role of socially inclusive spaces in the sustainable development of metropolitan regions → rather unclear	2 of 3 sub-objectives could be general INTERREG objectives	Overall objective does not allow a real idea of what was to be achieved	Unclear, no problems and potential results addressed
SEWAGE	To find the best way(s) for eliminating pharmaceutical substances from waste water → clear objective	Four sub-objectives clarified task; example: *development, testing, evaluation and monitoring and gaining operational experiences of treatment technologies and outcomes'*	Overall objective allows an idea of what was to be achieved	Clear research question, problems and potential results addressed

Source: by author, based on project applications

B Project Tasks and Methods

Objectives need to be further refined and translated into tasks, methods and responsibilities. A clear objective with concrete sub-objectives, broken down into specific tasks with clear responsibilities provides a distinct 'roadmap' that includes the logic of how single contributions add up to the final project result. The more complex a project, the more important it is to agree on which tasks and methods will lead to the projected results. Causal ambiguity and heterogeneity of tasks, on the other hand, negatively impact the effectiveness of experience accumulation, knowledge articulation and codification (Zollo and Winter 2001).

While objectives primarily provide foci for decision-making, *tasks* focus on action (Lundin and Söderblom 1995). An agreement on how to proceed from a present state to a desired state requires the planning of *working methods*. The ex-post evaluation of INTERREG III stresses the importance of transnational working methods for the use of learning options in a project (Panteia et al. 2010). For transnational knowledge transfer and development, the knowledge produced needs to be shared. To achieve truly transnational results, the implementation of concepts and strategies needs to be based on joint knowledge and carried out in a coordinated way. Table 5.3 summarises the projects' translation of objectives into tasks, the occupation with appropriate methods and their organisation in the four case studies.

The interplay of objectives, tasks and methods was quite coherent in the case of WOOD and SEWAGE. In RIVERS and PARKS, on the contrary, the process was more open-ended, objectives were not directly translated into tasks, and no concrete results were identified (for example the nature of recommendations). In both cases, this led to problems when producing common results. Although, in theory, an open-ended process can have its merits, in RIVERS and PARKS it constituted a substantial barrier to the production of transnational results.

Table 5.3 Tasks and methods in the case studies

	WOOD	RIVERS	PARKS	SEWAGE
Translation into tasks	Sub-objectives translated into work packages	No clear links between objectives and action plan	No action plan, no work packages	Sub-objectives translated into work packages
Methods to use	Events, excursions, brochures, certification	Pilot projects, evaluation (not done)	Pilot projects, transnational feedback groups, unclear 'demonstration' method	Mass flow, chemical and biological analyses, pilot projects, MCDA, LCA

Source: by author

The *structure and logic of the work plan* influences a project's chain of action. A comparison of RIVERS and SEWAGE illustrates how substantial the impact can be: in SEWAGE, work packages logically built on each other (analysis, testing, evaluating) and pilot projects came into play during all of these. The nature of the final result was clarified at the outset and based on all project steps. In RIVERS, there were no direct relations between individual work packages, which were extremely complex (some could consist of up to 28 sub-actions). It is questionable how well partners were able to keep an overview of the contribution of each action and their strategic fit.

Methods that provide substance to transnational exchange can support learning processes. A survey of learning in interregional projects shows that participants experienced thematic and comparative studies as well as study visits and seminars as supportive to joint learning processes (INTERREG IVC 2013). Additionally, evaluation and joint implementation assist joint knowledge development and learning, and specific exchange methods make use of the transfer potential. However, the survey also shows that these activities need to be connected in a logical way to contribute to learning.

With respect to methods, the case studies were characterised by large differences and in three case studies, the process of turning project objectives into results with the help of appropriate methods remained a 'black-box'. While WOOD followed an exploitation strategy, SEWAGE mainly pursued an exploration strategy. These strategies require appropriate methods, for example, facilitated exchange with a focus on transferable aspects for exploitation strategies or joint tasks for exploration strategies. Methods in SEWAGE were harmonised and substantially discussed during the project process.

> My impression was that it had not really been thought through in the project how [the transnational feedback sessions] would work in a targeted way. And how they could actually help the recipient of the consultation in their daily work. (Partner PARKS)

PARKS' working methods (transnational feedback sessions, thematic symposia) pursued both explorative and exploitative strategies, which, however, were not directly linked to project objectives. On closer examination, however, partners seemed to be unsure about the purpose of exploitative methods and how they could be used most beneficially (*see quotation above*).

RIVERS originally aimed at implementing pilot projects on a joint knowledge basis and letting these contribute to the development of new knowledge (guidebook). Partners were supposed to gather relevant existing knowledge in their regions, which would subsequently be compared (see Figure 5.2). This comparison would serve as the basis for knowledge abstraction, which again would be implemented in the pilot projects. The experience of pilot projects was to be evaluated and abstracted into a 'guidebook'. Finally, the guidebook would be validated with further experience and the model adapted. However, this working approach was quickly abandoned by the partnership due to a lack of translation into the action

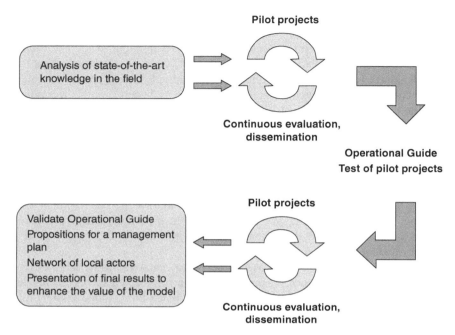

Figure 5.2 Project scheme RIVERS

Source: adapted from project application

plan, a lack of identification of project partners with the approach and the loss of the project coordinator (the author of the scheme) during the project.

While WOODS only worked with joint actions, in RIVERS partners did not find adequate methods for exchange and knowledge development.

C Partner Integration and Transnational Division of Labour

A project strategy requires the allocation of responsibilities (division of labour) and the integration of the various project partners into the project. In project reality, partners' integration in project components varies considerably due to the composition of partners, their range of interests and mixed contribution in terms of budget and manpower.

In their analysis of the impact of the structural configuration of social capital on the performance of temporary organisations, Di Vicenzo and Mascia (2008) identify *strongly interconnected networks* as more likely to develop trusting relationships between members, benefit from greater cooperation, greater conformity to norms, greater information sharing, less tendency to engage in competitive behaviours and an increased willingness of project members to engage in discussion and knowledge exchange. Other authors support this and also point to a higher innovation potential (Saxton 1997; Ayas and Zenuik 2001; Keller 1986). A high interaction rate eases the comprehension of the knowledge transferred and supports a 'common language' (Pérez-Nordtvedt et al. 2008). Moreover, cooperation intensity, depth and appropriate opportunities have been linked to projects' potential for learning, as has the existence of truly joint tasks (Panteia et al. 2010; Colomb 2007). People participating only in a few meetings, therefore, learn less than those who are more involved (Baird et al. 2014). Thus, the inclusion of project partners with their diverse interests and knowledge sources are important enabling factors for learning processes (Schusler et al. 2003; Moster et al. 2007).

In transnational projects, ties between partners are established by people using opportunities to meet, exchange and get to know each other and by working together on concrete tasks. 'Task interdependence' (Lubatkin et al. 2001) means that participants understand that their agenda is best realised when they specialise in the activities at which they are individually most competent and is a necessary ingredient in cooperative learning processes.

In a well-interconnected project, partners are integrated into the project structure and able to participate intensely in project tasks. In a study of learning in interregional projects, project partners identified low participation, a lack of continuity and an inadequate organisation of the learning process as particularly hampering factors for learning processes (INTERREG IVC 2013).

In the logic of transnational projects, *work packages* are potential ties between partners, which either follow a course-of-action logic (SEWAGE: analysis, solution, assessment) or a thematic logic (WOOD: certification, forest management, marketing). Except in PARKS, the projects assigned specific responsibilities to partners and work packages worked as a tool for creating ties between partners. In PARKS, partners met in thematic symposia and transnational feedback sessions. These seemed to compensate for a lack of transnational integration and division of labour and did not have a balanced participation by all project partners.

Table 5.4 depicts the different intensity of partner involvement in the WOOD project and provides an idea of potential ties in a project with 13 different actions. While some partners intensively cooperated, others only participated in few actions and thus faced difficulties in building cooperative relations and linking to the overall project. The design of work packages influenced how well partners with complementary expertise exchanged knowledge.

> The ministry . . . is the perfect example, they just send someone who would attend the meetings, but their input was next to zero. (Partner RIVERS)

> They contributed with their work at [place name], gave money for a brochure, but nicely kept out of all discussions. (Partner RIVERS)

Another example of unbalanced partner integration is RIVERS: a large group of partners only worked with their individual pilot projects (mainly public authorities) and remained somewhat unconnected. Only a small group was involved in common actions (NGOs). Interviews revealed a lack of trust between the two groups, which may be related to weak integration. This probably had a bearing

Table 5.4 Gradient of integration in WOOD work packages

Dominant partner	Strong involvement of some partners	Weaker involvement of some partners	Marginal involvement of some partners
Partner as work package leader for 2 of 3 work packages	Partner involved in 2 of 3 work packages	Partner involved in only one work package	Partners involved in only 2–3 of 13 actions
Partner involved in all work packages and all actions	Partners involved in 8–13 of 13 actions	Partners involved in 4–7 actions	Partner involved in only 1 action

◄───►

High involvement   Low involvement

Source: by author

on the exchange and cooperation processes during the project (Di Vicenzo and Mascia 2008; Saxton 1997), which – in theory – could have benefitted from the strong interdependencies between the two parties (see section 5.2.1).

As will be seen later, the projects with unbalanced partner integration and fewer chances to build close ties faced greater difficulties during the knowledge transfer and development process. While WOOD and SEWAGE had certain 'task interdependence' and a clear division of labour that had been intensely discussed during preparatory meetings, the lack of congruence between objectives and tasks in RIVERS left open questions about partner responsibilities, and knowledge processing remained unsolved and unassigned.

In transnational cooperation projects, partners meet in so-called 'partner meetings' (except in the PARKS project) to update on individual and group achievements. These support the establishment of ties, but some structures proved to be more helpful than others in the case studies: structured along work packages, they were more integrative than when being structured along pilot projects. In the latter case, meetings were challenged to establish links between partners.

D Pilot Project Set-Up

According to the INTERREG IVB programme for Northwest Europe, a 'pilot project' or 'transnational investment' has a clearer impact in other countries than the one where it is implemented and is of common benefit to the partnership. In the new programme, It can either be a joint investment that is developed and used by all partners, the decentralised component of a network of investments that only function as an integral entity or a 'replicate investment' of a jointly developed concept (INTERREG North-West Europe 2015). Still, opposed to work packages and joint actions, pilot projects are until today mostly implemented locally by one partner or a regional partner consortium and without transnational cooperation. Thus, they represent the 'non-transnational' element of cooperation projects and reduce their transnational dimension (Böhme et al. 2003). As they tie considerable costs and manpower and account for the practical project results, the set-up of pilot projects potentially strongly influence the overall knowledge development and learning process.

> My impression was that there were projects that were much more orientated towards implementation . . . and then those that started with studies, which may lay the basis for further action. (Partner RIVERS)

> The . . . conference, for example, was only a political measure, nothing more. It was simply not possible to get anything more out of it. (Partner RIVERS)

Today, most INTERREG B projects work with pilot projects (see Chapter 2). In the case studies, these served quite different purposes: some were highly research-driven and used pilot projects as laboratories (exploration), others contributed to innovation and knowledge generation by practically testing out project findings. A third group of pilot projects was used for communication and advocacy by aiming at raising awareness and 'demonstration' (Vreugdenhill et al. 2010) and did not have a knowledge function. The knowledge type produced by pilot projects had a bearing on the knowledge characteristics of the overall project and thus on the course of its knowledge process.

> The actions were very successful but were not really about public participation and more about PR. (Partner RIVERS)

> The main topic was clear. Public participation and the Water Framework Directive were the main aspects and below that level, lots of different things accumulated. (Partner RIVERS)

In theory, if pilot projects are supposed to demonstrate the viability of project results and how these – in all their components – work in practice, they need to be of a strategic fit by:

1 being able to demonstrate exactly what is to be shown by the project in practice, picking up its leading question or a particular aspect of it and thus clearly fitting to the project topic (thematic identity and vertical fit);
2 strategically fitting together, covering the various relevant aspects, complementing without repeating each other (horizontal fit);
3 containing the project's transitional aspect and the relevant development stage in relation to the implementing partner and/or region (identity of objectives and vertical fit).

While WOOD only involved joint actions, the other three case studies involved pilot projects of varying degrees of vertical and horizontal fit. Thematic identity was related to the overall subject or particular work packages and other components. The horizontal fit between pilot projects was created by comparability or complementarity, depending on their purpose. Thus, the strategic fit can be enhanced by a conscious planning of the pilot portfolio and taking into account partners' different development stages. In contrast, projects that serve as an umbrella for a range of local and unconnected projects face substantial difficulties in linking these to the project's overall objective and work packages, but interviews also disclosed that partners did not always see a need to clarify the contribution of pilot projects to project's objectives.

SEWAGE was able to achieve a horizontal fit of pilot projects by *comparability*. Relevant theoretical knowledge was limited, and pilot projects served as a transnational laboratory to test new approaches. The comparison of pilot projects derived the project findings. This was the only project where pilot projects had a direct knowledge-creating function for the overall project. Alternatively, pilot projects can be of a *complementary design*.

> Our pilot project emerged from a long wish list that had piled up in our organisation. . . . From this long list, we chose a project that fit, was feasible and had not yet been implemented. (Partner RIVERS)

In PARKS and RIVERS, unconnected local projects impacted negatively on the vertical and horizontal fit of pilot projects. These were neither directly comparable nor complementary.

> The British pilot project was very local and quite fragmented while we tried to realise a regional park over 80 km and real structural change. (Partner PARKS)

Within the same project, pilot projects could be of different sizes and scopes and pursue different objectives (*see quotation above*), which made comparisons and links between them problematic and generally increased the complexity of the knowledge produced at project level. In PARKS, for example, some pilot projects focused on participation at local level, others on spatial visioning at regional level or on creating 'regional parks'. Synergies were only used to a limited extent due to a lack of awareness and planning. This lack of strategic fit did not make exchange impossible, but limited partners' ability to draw a concrete benefit in terms of feedback and transfer options.

> We never sat together to discuss what we want to do, which measures would be comparable and if they could form clusters. (Partner PARKS)

Pilot projects also strongly differed with respect to their development phases. While, for some partners, their pilot project was the first contact with the topic at hand; others were well experienced and had a substantial knowledge base to draw from. Although the differences in development stages represent a potential for knowledge transfer, they need to be perceived as such to allow transfer (see section 5.2.1B). Different levels of previous expertise also had a bearing on the choice and design of pilot projects. While some pursued very innovative approaches, others had more moderate approaches. In RIVERS, for example, some partners limited public participation to public information, while others implemented highly inclusive role-plays and similar measures. Although pilot projects may be sometimes used for catching-up processes, their innovativeness affects the novelty and value of the experience for the overall project. Differing framework conditions and development stages can become barriers to knowledge transfer and development, but a lack of strategic fit between pilot projects increased cooperation barriers even more.

Project Strategies: Conclusion

PROJECT OBJECTIVES

Setting objectives is key to the project strategy and of multidimensional character. Project objectives have an individual and a common dimension as well as a task and a transitional dimension. A particular challenge of transnational cooperation is the potentially high diversity of interests and objectives of partners. It is important for transnational projects to have a certain congruence of objectives to provide a strong joint framework for strategy, cooperation process and results. Individual project objectives in WOOD and SEWAGE overlapped and were rather concrete. In contrast, in PARKS and RIVERS, which had less clear objectives, interviewees had less 'goal interdependence' and more individually dispersed objectives. Moreover, project objectives impact on the overall direction of the cooperation process and the ability to achieve joint results. A lack of joint objectives or identification with them reduced RIVERS and partly PARKS to umbrellas of individual pilot projects, hampered the perception of the projects as coherent entities and impacted on all other strategy-related parameters.

Project objectives are the starting point for defining relevant tasks and methods, a division of labour and the choice of pilot projects. The definition of clear objectives thus guides the overall project strategy. Although these findings seem self-evident, the case studies underline that joint and concrete objectives are not necessarily a given in transnational projects. Setting objectives is also connected to an awareness of causal relationships. In SEWAGE, it was helpful for achieving objectives, adopting measures and plans accordingly and for formulating transferable project findings and conclusions. The more complex a project and the less clear its objectives, the more difficult it is to define relevant causal relationships and thus to reach objectives.

Project objectives potentially influence:

- all other structural parameters: partner mix and strategic fit, knowledge types and characteristics, tasks and methods, division of labour, choice of pilot projects, transnational logic;
- process parameters: providing a direction to exchange and transfer, awareness of the necessity of knowledge processing, finding of crosscutting issues, systematisation and generalisation. The awareness of causal relationships of how to achieve projected results influences the choice of methods and is itself influenced by intense communication and possibly a focus towards joint conclusions. Finally, project results and learning effects can only be assessed in the case of clear and concrete objectives (see section 5.3).

PROJECT TASKS AND METHODS

In general, the 'black-box' between turning a set of project objectives into projected results was a challenge in more than one case and is a factor given inadequate attention in INTERREG programme discussions. The coherence of objectives, tasks and methods varied considerably between the cases, as did the coherence and logic of the work plans. 'Exploitation' and 'exploration' strategies not only require a certain level of diversity in knowledge and experiences, but also a translation into concrete tasks. This can include focused exchange and feedback sessions (exploitation) or cooperative tasks (exploration).

Moreover, there were large differences with respect to methods, which potentially provide projects with a 'roadmap' for knowledge development. Particularly knowledge-relevant tasks require transnational methods to produce transnational insights and results. As seen during the participant observation, streamlined and standardised methods increase the transferability and generalisability of project findings but may not be feasible in every project. The other case studies illustrate how insufficient attention to causal relationships and working methods severely challenged the deduction transnational results.

Overall, it can be said that the clear project objectives in WOOD and SEWAGE provided reliable 'roadmaps' for project action, which – although this differed in intensity – is also reflected in the methods. The high level of expertise among partners may also be a possible reason for more sophisticated working structures. In contrast, PARKS made no particular efforts to open up the 'black box' between setting objectives and achieving results while RIVERS abandoned its project methodology. This led to a considerable degree of confusion on how to develop joint knowledge and to the production of project results that not all partners identified with.

Project methods potentially influence:

- other structural parameters: knowledge types (for example validity of knowledge) and division of labour; if pilot projects are given a task with respect to the overall project (for example provide data), also the choice, design and 'strategic fit' of pilot projects are affected;
- process parameters: intensity of knowledge exchange (internal working logic); commonalities between partners and thereby the transferability of experiences and knowledge; knowledge processing strongly depends on a relevant framework of methods (see section 5.3).

PARTNER INTEGRATION AND TRANSNATIONAL DIVISION OF LABOUR

The variety of partners involved in a transnational project makes the development of close ties a prerequisite for cooperation and knowledge sharing and a fruitful learning process. These ties are established by making use of opportunities to meet and exchange, integrating into the project structure and working closely together to achieve specific tasks. In the case studies, the integration of partners into project actions and their division of labour made a substantial difference to the transnational character of the projects and their options for joint knowledge development.

Work packages relate to the overall project and each other either thematically or functionally. Thematic work packages risk pulling up topic-related boundaries while functional work packages seemed more likely to invoke a division of labour. The organisation of meetings along individual pilot projects, work packages and cross-cutting issues that support inter-related thinking determined how well partners exchanged and cooperated.

As the four case studies show, the degree of partner integration and thus of cooperation intensity and depth varies considerably. Less integrated partners obviously face more difficulties to build up ties and thus trustful relationships with other partners, but also to make the best use of the transnational transfer potential. Based on the case studies, a direct link between partner integration and their contribution in producing transnational project results can be assumed.

Partner integration can be influenced by:

- partners' objectives: a focus on pilot projects limits the involvement in actions and work packages and the transnational potential. A focus on joint products and processes is more likely to induce an interest for participation in cooperative tasks and actions.

Partner integration potentially influences:

- structural factors: transnationality (existence of common actions, cooperation structures);
- process factors: intensity of communication, exchange of information and experiences, feedback, accessibility and comprehensibility of knowledge and thus the transferability of experience as well as the ability to produce joint results (see section 5.3).

PILOT PROJECT SET-UP

Pilot projects are supposed to demonstrate the viability of project results and how these work in practice. The high share of pilot projects in the budget and time effort of transnational projects grants them an important role and induces a substantial impact on cooperation and knowledge development options. However, the concept of pilot projects has not yet been well conceptualised in INTERREG programmes. There is an unsolved conflict between pilot projects delivering 'tangible' results and the risk of reducing projects' transnationality by focusing on individual actions that mainly benefit individual regions or organisations. Neither has the fact that pilot projects can have diverging objectives and serve different purposes as shown by the case studies been given much attention. Thus, the conception of a strategic fit of the overall project focus and its pilot project portfolio – especially in respect to development stages – is still missing.

Pilot projects had varying degrees of vertical and horizontal fit. Different development stages pose a potential for transfer and exchange but need to be consciously explored and exploited. The innovativeness of pilot projects affected the validity and value of the experience for the overall project; links between them impacted on the relatedness of experiences made, while their design and content influenced the complexity of the knowledge involved. Different sizes and scales of pilot projects further complicated comparisons and inter-linkages.

Comparability, complementarity, synergies and a cohesive use of data support the production of joint project results and can overcome the fact that pilot projects often represent the non-transnational element of transnational projects. While proper planning supported the strategic fit of pilot projects in SEWAGE, a rather coincidental choice of pilot projects decreased both the horizontal and vertical fit in the other two cases. The lack of a basic strategic fit of pilot projects indicates that far too little is understood about their potential and the impact of their existing or lacking co-action on joint knowledge development. As will be seen later, their fit or non-fit decisively influenced the projects' ability to transfer and develop knowledge.

The set-up of pilot projects is potentially influenced by:

- partners' organisational type, objectives, tasks, knowledge types, 'development stages' with respect to the topic, partners' strategic fit.

The set-up of pilot projects potentially influences:

- knowledge characteristics: validity and value of experience made, knowledge complexity, relatedness of the experiences and the transnationality of the overall project (see section 5.2.2);
- process factors: knowledge input, usefulness and intensity of exchange on pilot projects, options for feedback, transferability of experiences (see next section).

5.3　From Individual Experience to Collective Knowledge: Process Parameters

Traditional approaches to project management and assessment of project performance focus on structural factors and do not usually take process and contextual factors into consideration. Although project structures are highly relevant as they lay the overall basis for the following process, which is then likely to follow a certain path, within a given project structure processes can still run very differently. Moreover, projects with various participants also include multiple learning processes. Only very recently have researchers started to pay more attention to the process-related performance of projects. These consider the impact factors such as knowledge sharing, communication, interaction patterns, coordination, and shared knowledge may have on project performance.

The following section attempts to open up the 'black-box' between the in- and output of knowledge in transnational projects with the help of some distinct 'process phases' relevant for transnational cooperation. These phases can only be an approximation to the phenomenon of transnational knowledge transfer and development as the variety and complexity of transnational project processes and the lack of theoretical conceptualisation do not allow the exhaustive coverage of all relevant process aspects and their exact mode of action.

The process phases are built on the above-discussed theoretical considerations based on experiential learning, knowledge creation and inter-organisational learning that help to understand the general exchange in transnational projects (Kolb 1984; Nonaka and Takeuchi 1995; Prencipe and Tell 2001; Huelsmann et al. 2005). Additionally, they are based on policy transfer literature that helps to shed light on the transfer of existing knowledge and the concept of 'observational learning' that contributes to the understanding of indirect learning. The process phases deducted in section 4.3 were amended with information and clues from the four case studies as to the relevance of these phases, significant challenges encountered and ways to deal with these.

According to the process phases developed in section 4.3, the process of transnational knowledge development and learning can be described as potentially passing through the following steps (see also Figure 5.3):

PHASE 1: Project partners make *new experience in pilot projects*, which allows the creation of new experiential knowledge.

PHASE 2: *Knowledge and experiences are exchanged*, amended and further developed during meetings, study visits, studies and reports and other forms of group work. This exchange can be more or less intense and focus on existing and developing knowledge, theoretical knowledge and practical experience and thus be a more or less appropriate basis for the project's knowledge transfer and development.

PHASE 2A: The *transfer of existing knowledge* requires the transnational cooperation process to start with a certain knowledge input, which is the combination of different knowledge bases (internal and external). Although not every transnational project involves processes of knowledge transfer between partners, they are part of the raison d'être of INTERREG B projects as they avoid the 're-inventing the wheel'-syndrome. This transfer is mostly a bilateral process where knowledge and experience existent in one place is of use to partners in another place.

PHASE 2B: Exchanging newly emerging *experiences from pilot projects* is related to integrating experiential knowledge into the overall knowledge development process. This is an important source for the quality and coherence of the overall learning process and project results.

PHASE 3: *Reflection processes* are required to process new experiences into knowledge. In this respect, a particular potential lies in making use of the transnational knowledge pool for common reflection processes. Transnational cooperation projects lend themselves to mutual feedback on individual pilot projects and cross-fertilisation allows new common knowledge to develop. Pilot projects gain from the variety of knowledge sources and joint reflection of experience. These feedback processes can be institutionalised or take place randomly.

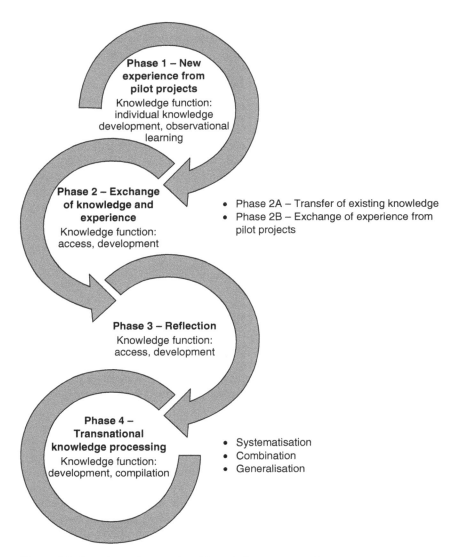

Figure 5.3 Overview of transnational knowledge processes phases

Source: by author

PHASE 4: Towards the end of the project, the transnational partnership has to produce joint project results, which need to go beyond the result of local projects. This requires a process of *transnational knowledge processing* that breaks through the borders of individual pilot projects, combines the different inputs of existing and emerging knowledge and makes sense of them in an integrated way to form results that are of interest and relevance to the whole partnership and possibly beyond. Transnational knowledge processing can again be divided up into three distinct components:

- the systematisation of the various individual knowledge inputs and outputs as,
- the combination of individual knowledge strands, and
- the generalisation and abstraction of transferable knowledge.

In the following, the case studies are used to gain more information on the real-life functioning of projects and challenges encountered.

5.3.1 PHASE 1: Making Experience in Pilot Projects

Phase 1 is highly individualised between projects, where project partners implement their pilot projects and joint actions and create the basis of new experiences that will then feed the subsequent process. Although pilot projects are supposed to take up relevant project findings, they are often planned in advance and thus mainly based on the knowledge base of individual partners before the project start. Besides planning their pilot projects individually, project partners usually also implement them individually. This initially limits their knowledge function to the production of individual and organisational experiential knowledge. With the evolution of the transnational programmes, programme authorities started to realise that pilot projects usually constituted the non-transnational element of transnational projects. Programmes after 2006 thus stress that pilot projects have to be of common benefit to the partnership and, therefore, need to be jointly designed, implemented, evaluated and used by several or all partners (see for example NWE Guidance Note on Transnationality[3]). In the reality of programmes, however, this has not been systematically assessed and, as can be seen in Chapter 6, in the NWE programme area, this requirement can often only be found in joint evaluation schemes for individual pilot projects.

5.3.2 PHASE 2: Transnational Exchange and Communication

The main purpose of transnational projects is to ensure the exchange of experience and knowledge between its participants that leads the project to achieve its objectives and accomplishes its transnational mandate. Although exchange is inherent in cooperation, it can pursue different goals and be of different intensity, which is often influenced by the individual partner motivation. For successful exchange, partners need sufficient motivation and reason to exchange – such as a common objective – and a trigger, such as dissimilar knowledge. This aspect is discussed in part A, below.

Additionally, the direction and intensity of exchange are influenced by the existence of appropriate structures, platforms and occasions to meet and discuss, a 'space' of social interaction (Nonaka and Takeuchi 1995) as well as sufficient time. Partners in transnational projects regularly meet for to update on individual progress and project management issues, to pursue joint tasks and to exchange

---

3 http://4b.nweurope.eu/index.php?act=page&page_on=documents&id=384 (accessed on 5 March 2015).

on the project topic. These partner meetings support both the access to existing knowledge and the development of new knowledge. Openness, transparency, an inclusive atmosphere as well as the sheer number of interaction moments enhance opportunities for all participants to learn and contribute knowledge (Mostert et al. 2007; Schusler et al. 2003). Particularly the integration of interdisciplinary teams benefits from ample time, including informal time, to exchange (Bruce et al. 2004: 468). Moreover, group-learning processes call for freely accessible and transferable knowledge (Capello 1999; Kissling-Näf and Knoepfel 1998). Exchange structures are discussed in part B.

Project communication is decisive for the course of all kinds of projects but faces specific challenges in transnational cooperation. Partner diversity tends to lead to higher complexity and thus poses a challenge to transnational projects, in particular, but it also represents a potential benefit. Communication and diversity are discussed in parts C and D.

A Motivation to Exchange

> There was no overall exchange of experience. There were many bilateral discussions, and these were very fruitful. However, they only took place between NGOs. (Partner RIVERS)

For transnational exchange to be effective, project partners require sufficient motivation, reason and a trigger for exchange. In SEWAGE, partners were highly interested in learning from each other and particularly from one highly experienced partner. In contrast, in RIVERS, not all partners were equally willing to open up to a transnational exchange of knowledge and experience. Partners from public administrations were described as less involved, less open to new ideas and more focused on their individual pilot projects (*see quotation above*). In PARKS, project partners were generally very interested in the exchange, but in terms of using the experience and knowledge of other partners, some found that only the experience from partners from the same country would be useful for them due to comparable framework conditions. Particularly the experience of the English partners was assessed as 'too different' and of little relevance to other regions. Thus, links between partners were somewhat unbalanced and often only bilateral. As opposed to RIVERS, where the exchange was limited due to a lack of openness and participation, in PARKS it was constrained by a lack of perceived benefit and awareness of one's own potential knowledge roles. In both projects, the lack of a strategic pilot project portfolio and a subsequent lack of relatedness and novelty of knowledge may have additionally decreased the partners' interest and motivation for exchange.

B Exchange Structures

For information and knowledge to flow, project partners engage in various ways of exchange. This exchange either happens coincidentally or is supported and facilitated by dedicated meeting platforms and joint tasks (see section 5.2.3).

Maybe we are . . . in a better situation than other partners, who only took part in transnational feedback sessions and got to see less of the overall project. . . . When you only participated in the feedback sessions you only got a taste of the process, and it was certainly difficult to see the whole project. (Partner PARKS)

- In all four case studies, *study visits* offered good options for exchange and discussion, if mostly bilateral.
- In RIVERS, partners met in *partner meetings* and a few small groups formed for specific tasks, but the project did not provide further support to its exchange processes. True exchange thus remained more or less limited to a smaller circle of partners.
- Although *work packages* do not guarantee efficient exchange, they break the project into different tasks that require exchange and cooperation. In SEWAGE, they followed a process logic, in which partners became subsequently involved. This established a division of labour, where the inter-linkages between partners necessitated exchange (see section 5.2.3). In WOOD, work packages followed a thematic logic, where partners split up in parallel working groups (which limited exchange). In both projects, project partners seemed content with their exchange and their ability to make use of partners' knowledge stocks.
- Partners in PARKS did not work with work packages and did not meet in partner meetings but in specific *transnational feedback sessions* (*see quotation above*) and *thematic symposia*. Here (and similarly in another project, where the author visited a partner meeting), discussions focused on one specific pilot project or topic at a time. Although discussions were intense, there was no exchange between individual pilot projects or topics.

Both in RIVERS and PARKS partners seemed less content with the project's knowledge exchange and had difficulties pointing to how they had benefited from the project.

C Project Communication

Interpersonal communication is highly important in transnational cooperation, as projects emphasise the people-to-people approach by meeting and interacting. People-to-document approaches become relevant during later project stages when results are retained for dissemination (Hansen et al. 1999). In the following, some general findings on communication in the four case studies, its focus and intensity, barriers encountered and approaches to communication management are discussed.

The importance of communication for successful project management has been well documented in project management literature (for example Adenfelt 2010; Pinto and Pinto 1990), but it is of particular importance to social learning. Communication is both an element and a precondition for knowledge sharing and transfer and plays a role in different models of knowledge creation (for example Nonaka and Takeuchi 1995). In particular, constructivist and situative learning theories emphasise the role of dialogue and communication for learning processes (for example Gergen 1995; Lave and Wenger 1991). Particularly in projects featuring a high degree of tacit knowledge, communication is at the heart of knowledge processes between the carriers of knowledge (Hansen et al. 1999). The basic function of communication is the flow of information, itself an outcome of the chosen communication tools as well as the characteristics of the communication process (Adenfelt 2010). The objective of transnational communication is to ensure this flow of knowledge, to share individual knowledge and to create new knowledge together. Communication also goes beyond the immediate project and extends to continuous links and feedback with the outside world, such as links to political decision-makers.

In general, a retrospective assessment of project communication is highly challenging, as interviewees' perception of communication processes is rather subjective. Moreover, to discuss and describe project communication requires interviewees with certain meta-awareness and the ability to reflect on the overall process. Although the interviewed project partners had difficulties reflecting comprehensively on project communication, much valuable information could indirectly be collected while discussing other aspects. The participant observation allowed deeper insights into communication aspects.

Some overall observations include a general lack of communication in RIVERS that worked as a barrier to knowledge transfer. PARKS was the only case where partners reported that culturally different communication styles had led to problems, and where ambiguity and interferences (when words sound similar but mean different things) about terms were high.

OBJECT AND INTENSITY OF COMMUNICATION

The character and intensity of communication depend on the number and relevance of ties between project partners and on the exact type of activities partners are involved in and thus communicate on. Here again, the set-up of pilot projects and communicative options that arise between them, a potential division of labour as well as discussion culture and trust played a role for the intensity of communication in the four case studies. Focusing on individual pilot projects set in very specific framework conditions, partners faced larger communicative challenges when trying to bridge the variety of individual experience as seen in the cases of RIVERS and PARKS. When partners worked in a division of labour on the joint development of new structures and processes as in WOOD and SEWAGE, they focused on crosscutting and common issues and regional and national differences had less of a disruptive function. Regional and national differences, although still being important, then have less of a disruptive function. Although it is not impossible to achieve effective communication in a project with a strong focus on pilot projects, it requires more effort to overcome the related barriers. In SEWAGE, where pilot projects were of a high comparability, partners engaged in detailed and targeted discussions during and after visiting pilot projects. Similarly, during a visit to another project with highly diverging pilot projects, fewer questions were asked, and discussions remained less intense.

Project communication develops and changes over time and is thus a process in itself. Often, it intensified with time in both quantity and quality, for example from a more general exchange on the topic to more targeted exchange in smaller work groups.

In transnational settings, communication is likely to encounter *language and cultural barriers*, possibly giving rise to ambiguity or even misunderstandings, but also to different mentalities with respect to dealing with problems and conflicts.

> He had a very polished and eloquent way of speaking English, talked very quietly and in very long sentences. [Partner] in contrast had a very clear and direct language. This mix did not work out very well. (Partner PARKS)

In terms of language, Hambrick et al. (1998) expect a low degree of common language facility to impair group functioning by hampering the exchange of information and trust. Increases in shared language facility, they argue, result in corresponding increases in group performance. In his book on intercultural management, Holden (2002: 299) identified three sources of potential 'noise' in inter-cultural communication: ambiguity, interference (when words sound similar but mean different things) and lack of equivalence. A survey among project partners in interregional projects shows that communication problems related to language were perceived as the main hampering factor for both individual and collective learning (INTERREG IVC 2013).

> In general, we discussed a lot. With 13 partners, we sometimes ended up with 13 opinions. These different points of view made the project exciting. You also notice different mentalities. (Lead Partner RIVERS)

> Sometimes the opinions about how to manage a forest differed, but they did not contradict each other. When this happened, we tried to find a kind of compromise. I am sure partners have learned a lot from this and how others work. (Lead Partner WOOD)

> Partners had very different understandings of public participation. For some, it was information of the public, for others, the public had to be actively involved. We had a very constructive exchange on this. (Partner RIVERS)

All four case studies had to put effort into dealing with transnational diversity. To benefit from diversity often proved highly challenging, and misunderstandings could not always be solved. However, willingness to compromise and sacrifice, flexibility and tolerance of ambiguity and a common vision seemed to help in overcoming differences. The effects of diversity were quite complex: while a high degree of cognitive diversity was beneficial to partners in most situations, behavioural diversity (such as different working methods, ways to communicate) mainly disturbed cooperation. *Cognitive diversity* arose from national, institutional or even personal differences, such as different viewpoints and approaches. *Diversity of values* proved a benefit to some partners in the projects dealing with public participation, particularly for those who dealt with participation for the first time. Still – owing to a lack of openness to alternatives and of communication – the potential posed by both cognitive and value diversity was not always fully used.

> I do not think our ministry was able to learn a thing or two from the Dutch partners. When I told them about the Water Framework Game, they stared at me bluntly. I think that this is also about different mentalities, it is easier to win over Dutch people for games like that. (Partner RIVERS)

In RIVERS, in the case of a particular instrument for public participation used by the Dutch partners, different framework conditions in Germany proved to be an obstacle to its transfer. A general lack of discussion and of awareness of relevant differences further impeded transfer attempts ('implementation stickiness'), possibly linked to a lack of motivation on the recipient side (see section 4.3.2). As a result, the projected transfer of the instrument from the Netherlands to other countries had to be given up. This points to a lack of 'absorptive capacity' (Cohen and Levinthal 1990) as the received knowledge was too different from the other organisations' mental representations and subsequently treated as something unique and not taken seriously.

High levels of diversity can increase a feeling of 'foreignness', but also present a potential for innovation and out-of-the-box thinking. This, however, needs to be used wisely, and the case of RIVERS illustrates how important the articulation and discussion of relevant differences is to be beneficial to partners. Diversity of framework conditions had a strong impact on the transferability of knowledge, particularly when this was strongly context-dependent. Low diversity, on the other hand, led to a feeling of 'closeness' in WOOD, where partners had a similar cultural background. A comparable closeness can also develop when partners have a similar professional background as in SEWAGE. This case illustrates that even when cognitive diversity is rather low, partners may still need to build up transnational collaborative know-how due to differing objectives, organisational and cultural backgrounds, different working paces and high levels of ambiguity. Then again, low institutional and professional diversity led to a situation where a lack of partners with insights into the bottlenecks of practical implementation proved to be a challenge for the 'translation' of the project findings to decision-makers. In the case of different 'development stages' among partners, the adaptation to more innovative approaches or higher standards was desirable to partners and diversity turned into a benefit.

Language issues in cooperation also include differences that stem from different disciplines or organisational backgrounds. With his concept of 'trading zones', Galison (1997) shows that groups of various disciplines can coordinate their approaches despite having very different approaches in instruments and characteristic forms of argumentation. They do this by finding an intermediate language that allows communication across disciplines. As Galison notes, these are situations that help innovation to occur. Similarly, in the planning discipline, communicative planning theorists have been concerned with mutual understanding. Healey (1992), for example, doubts that in multicultural planning truly shared comprehension can be achieved. Instead, cooperation partners should invest as much as possible in mutual understanding, but also be aware of what is not understood.

> I feel we are trying to re-invent the wheel or rather many wheels in all those places. In my daily work I exactly know what to do, I never have situations like this. (Partner SEWAGE)

During the participant observation, it was possible to observe communication processes while they were taking place. In SEWAGE, partners' perception and handling of communicative challenges changed over time, as they grew more accustomed to transnational cooperation. During early meetings, discussions were of a complex and fundamental character but lacked a common basis. Some partners were left frustrated about long decision-making processes and a general ambiguity of meanings, particularly those with little transnational collaborative know-how (*see quotation above*). Later on, these problems were more or less overcome; partners knew each other better, were more used to transnational communication and had found ways to relate to each other. When misunderstandings arose about the last work package towards the end of the project, partners already had a history of cooperation and were able to resolve the lack of clarity rather quickly. Also translational and visual communication tools can help to support a common understanding.

In PARKS, by and large, *interference* (words meaning different things) was particularly high: project partners spent a lot of time discussing terms such as 'participation', 'inclusive planning' or 'regional planning'.

COMMUNICATION CULTURE AND MANAGEMENT

Research on interdisciplinary collaborations shows that bringing people together and coordinating their conversations is not enough to ensure an efficient and result-oriented communication flow. Facilitation and mediation are required to collectively define objectives, strategies and outputs and to extract all relevant knowledge from participants (Desprès et al. 2004). In this respect, sufficient platforms, communication structures and discussion culture are of relevance. Communication structures such as meeting agendas, background papers, facilitators and the frequency and character of meetings can guide a communicative process (Knippschild 2008). Facilitation can also help for inter-connected thinking with questions that make partners think beyond their individual pilot projects. In this respect, the transnational feedback groups in PARKS are a good example (see section 5.3.5).

> I do not even know the real motivation of other partners [to participate]. (Partner RIVERS)

During several project meetings and study visits, the author was able to witness how much of a bearing project discussion cultures had on the general knowledge exchange. In SEWAGE, the character of meetings was positively influenced by an open atmosphere, in which all partners lay open their motivation and objectives. As the quote above shows, this is not always the case. To nurture its discussion culture, the Lead Partner of this project actively invited particularly experienced partners and members of its Scientific Board to provide feedback, summarised discussions and made sure that conclusions had the backing of all partners. This often triggered further discussions. Moreover, partners had a strong commitment to identifying with the project's findings. While SEWAGE was able to establish an intense and engaged discussion culture, partners in two other projects (see section 5.1) were much less able to make connections between individual presentations of pilot projects. During their meetings, communication was less oriented towards a real exchange and common cooperation objectives played less of a role.

> I sent [the final recommendations] round so that partners could comment. However, people were just glad that someone had taken on the task and – with a few exceptions – there were not many discussions. (Partner RIVERS)

This was similar in both RIVERS and PARKS. Consequently, project discussions were less able to create links between partner experiences and project results. An important conclusion is thus that targeted communication is related to a project's expected results and conclusions.

5.3.3 PHASE 2A: Exchange of Existing Knowledge

A Creating a Common Knowledge Base

Partners' existing knowledge stocks are the main knowledge contributors to transnational knowledge development. This emphasises the *strategic choice of the transnational partnership*, which was not always given in the case studies (see section 5.2.1). Additionally, projects can make use of external experts. Knowledge input can be of abstract, factual, tacit, explicit or experiential nature, which potentially influences the process of knowledge development and learning.

> In other projects, we try to find out about partners' competencies and how these can be best made use of. We find it important to see if people have something to contribute and to provide appropriate platforms to share these contributions. However, this was not done in PARKS; there was never the question about partners' specific expertise. (Partner PARKS)

To access their existing knowledge sources, partners need to activate and cross-link their knowledge and relate expertise (Lullies et al. 1993). Projects can incorporate specific tools for a systematic *collection of existing internal and external knowledge*, such as state-of-the-art-reports, literature studies or mapping exercises of potential knowledge sources (see Chapter 6). These tools bring knowledge together in a codified form but require good project management and facilitation with interdisciplinary skills. Knowledge combination is a time-consuming task that depends on an understanding of the potential contribution of the various represented disciplines (Bruce et al. 2004).

Three of the case studies worked with instruments, which created an overview of relevant existing knowledge. In SEWAGE, partners' initial skills and expertise were mapped to help the identification of potential knowledge 'senders' and 'receivers' and to assign responsibilities. This was not a theoretical exercise but necessary due to shared tasks. In RIVERS and WOOD, small groups of project partners compiled existing internal and external knowledge on selected topics. In RIVERS, however, this knowledge could not be used as an information basis to the project process due to a late completion.

PARKS did not compile codified existing knowledge but invited a broad spectrum of external experts to several thematic symposia and thus included their knowledge en route.

> The study on public participation . . . was not completed as early as anticipated. This was because project partners did not return the relevant questionnaire promptly. Furthermore, the information available on the subject proved to be more extensive than expected. (Activity Report RIVERS)

Ideally, pilot projects build on the compilation of existing transnational knowledge. In reality, programme requirements that usually do not leave much room for flexibility and that by and large require projects to define the relevant process elements up-front challenge this course of action. Expecting projects to complete their knowledge collection during the preparation phase is not realistic either. In effect – and as happened in three of the case studies – projects that spend time collecting existing knowledge are likely to start the practical project process in parallel, although these two processes would ideally build on each other. In SEWAGE, RIVERS and WOOD the time consumption of knowledge collection was underestimated. In RIVERS, knowledge collection took place in parallel with the implementation of joint actions and pilot projects and was finalised two years into the project. The methods applied in pilot projects could thus not build on a joint knowledge base as intended, and the latter could only contribute to the final report (*see quotation above*). In SEWAGE, on the other hand, partners agreed to postpone other activities until the finalisation of the literature study to ensure the integration of relevant insights into pilot project design.

B Transfer of Knowledge and Practice

When transnational project partners learn from each other by moving existing knowledge, one can speak of transnational knowledge transfer, which mainly has a knowledge access function. The basis for knowledge transfer comprises different areas of expertise, a knowledge gradient and the perception of analogous 'knowledge roles' as senders or receivers (see section 5.2.1), which can place both via articulation and codification (Mason and Leek 2008). It refers to whole concepts, strategies or instruments, rather than the exchange of pieces of knowledge.

The following section discusses how much project partners in the case studies were able to increase the benefit of transnational cooperation by making use of its transfer potential. It looks into the types of transfer objects encountered and what these meant for the transnational transferability of knowledge. Moreover, it analyses the type of transfer processes that took place and possible reasons for encountered transfer barriers.

TRANSFER OBJECTS AND FRAMEWORK CONDITIONS

With the thematic spectrum of transnational INTERREG programmes ranging from innovation over transport to resource efficiency and low carbon strategies, the cooperation objects of many projects are characterised by a high degree of context-dependency. For knowledge transfer, this implies that potential transfer objects are not similarly transferable; but that essential elements are linked to institutional legacies, state traditions or the dominant legal culture of the countries involved. There is no certainty that practices transplanted to another country deliver similar success as in their source country.

To understand the factors that impact on transferability, the literature on policy transfer and inter-organisational learning provides valuable insights (see section 4.3.2). Most importantly, the type of 'transfer object' has a bearing on the transferability of knowledge. These 'objects' can be institutions, policies, procedures, ideas, attitudes, ideologies or justifications. Transfer objects of the constitutional level and of low visibility (ideas, principles of action and philosophies) and high visibility (programmes, mode of organisation and institutions) are particularly challenging (OECD 2001). In contrast, transfer objects of the operational level and of medium visibility such as technologies, operating rules and know-how have the greatest potential to be transferred. Moreover, the characteristics of knowledge seekers and providers, the nature of their interrelationship (such as institutional similarities) as well as knowledge-specific variables (such as tacitness, ambiguity, complexity and specificity of knowledge) have been identified as influences on knowledge transferability (see section 3.2).

Although extensive transfer processes could not be made out in the four case studies, a few attempts to transfer components at operational level and of medium visibility could be found. In the case of very concrete and practical transfer objects, such as PR measures in WOOD, or standardised instruments, such as a wood certificate in WOOD or a filtering technique in SEWAGE, interviewees assessed these attempts to have been successful.

Contrasting WOOD and SEWAGE on the one hand with RIVERS and PARKS on the other demonstrates the relative ease of knowledge transfer in cases of natural science subjects and technical aspects due to their low context-dependency and options for verification. Large differences at the constitutional level and a high share of context-dependent knowledge made knowledge transfer attempts in RIVERS and PARKS less successful. Context-dependency again increased ambiguity and complexity and was additionally raised by the fact that project partners were practitioners and mainly gained experiential knowledge in pilot projects. Potential transfer processes were therefore on secondary experiences. Their person-bound character and context-dependency need to be reflected and their re-use is limited to similar contexts or is dependent on the advancement to a level of higher generality through processes of de-contextualisation (see section 5.3.6). Meanwhile, the high specificity of individual pilot projects further limited transferability, whereas the low specificity of the project objective in both projects meant that they lacked overall focus. Particularly in RIVERS, partners did little to identify potential 'knowledge roles' and existing knowledge stocks were not used for transnational transfer. Thus, knowledge accessibility did not guarantee acquisition (Inkpen 2000).

OUTCOME OF KNOWLEDGE TRANSFER PROCESSES

> During the workshop, some partners wanted to sell a Dutch method as *the* method for public participation, but that turned out to be a flop. For the German context, this method was completely useless. . . . You simply cannot do public participation by the numbers. (Partner RIVERS)

A general impression from the four case studies is that although some knowledge transfer took place and led to 'inspiration' processes, these were difficult to track down in more detail.

For project partners in WOOD and SEWAGE, knowledge transfer was an integral part of the project and motivation to participate. It led to the *copying* of practical and technical instruments, approaches and solutions. In RIVERS and PARKS, interviewees did not think direct copying to be possible, also because of the different development phases of pilot projects. In RIVERS, attempts were made to transfer an innovative instrument for public participation from the Netherlands to Germany. However, the potential receiver did not spend much time to familiarise with the transfer object and its framework conditions (*see quotation above*). Other forms of knowledge transfer such as *emulation* were not considered. This 'implementation stickiness' (see section 4.3.2) was caused by communication problems with coding schemes and cultural differences and a lack of know-how and know-why about the transfer object. Such insufficient attention to context and reduction of complexity is also called 'inappropriate transfer' (Dolowitz and Marsh 2000). As comparative research shows, this is a common situation in international knowledge transfer when the receiving side lacks belief and involvement in the process and fails to adjust the knowledge to its own context (Kroesen et al. 2007).

> I would have wished for a stronger linking of regional projects to the main project questions. Like a matrix, where you can see which topics in which projects could be followed up with specific guiding questions. (Partner PARKS)

In PARKS, the multiplicity of options for observational learning in 'transnational feedback sessions' led to knowledge transfer processes. The unusually long project lifetime of six years made room to familiarise with the different contexts of transfer objects and to partly overcome potential transfer barriers. In the case of the 'regional park' concept, some emulation processes took place (see section 5.2.1). Beyond this example, however, the lack of strategic fit of pilot projects posed a critical challenge (*see quotation above*).

In both RIVERS and PARKS, it was striking how quickly transfer options were abandoned as soon as the first challenges emerged, which may point to a limited 'absorptive capacity' of project partners (Cohen and Levinthal 1990) or the 'stickiness' of the knowledge involved (Szulanski 1996). In general, the inability to recognise opportunities for transfer can be found in RIVERS, PARKS and two additional projects the author visited. They all had pilot project portfolios of little strategic fit and a lack of a joint conceptual basis. As both aspects impact on the project focus and purpose, a lesson from these cases it that projects of this type are likely to encounter knowledge transfer barriers and 'initiation stickiness' (see section 4.3.2).

5.3.4 PHASE 2B: Exchange on Pilot Projects

The fact that project partners usually plan and implement their pilot projects individually, initially limits their knowledge function to the production of experiential knowledge at individual and organisational level. However, if pilot projects are to produce explicit, transnational and transferable knowledge, projects need to ensure that experience is translated into knowledge, a process linked to reflection (see section 4.3.3). What challenges the projects in the case studies encountered when trying to make use of their pilot projects as a source for knowledge development and how these may have been overcome is discussed in this section.

All projects except WOOD worked with pilot projects. These tended to take up large parts of the project budget and time effort. They constituted the one element that all partners were involved in and can thus be understood as creating the most relevant potential for exchange and transfer. The implementation of pilot projects allows partners to make new experiences and learn locally and closely related to existing routines. This process is also called 'experience accumulation' and can take place either through learning-by-doing or learning-by-using, which are both related to single-loop learning (Prencipe and Tell 2001). Ultimately, experience accumulation can also be the basis for the development of new knowledge.

The particular potential of a transnational project lies in the fact that partners make different practical experiences at the same time, which allows for observational learning (Bandura 1979). In principle, exchanging the experience from pilot projects can either refer (1) to comparing different approaches and strategies to similar problems or (2) to gaining practical implementation experience (what works where, how and why). Observational learning can stimulate creativity by exposing observers to a variety of models that causes them to adopt combinations of characteristics or styles (adaptive learning). It allows learning about unconventional strategies when observing unconventional responses to common situations (generative learning). The added value for cooperation partners lies in the variety of approaches that can both stimulate their own work and permit them to draw lessons from more than one's own case (learning *from* each other). In addition to using the experience of single pilot projects at an individual level, transnational cooperation also enables joint sense making of a whole portfolio of experiences with the help of diverse cognitive frames for discussion, reflection and abstraction (learning *with* each other). The added value of pilot projects is then based on several cognitive frames reflecting on several cases instead of one cognitive frame reflecting on only one case. Experiences that others make can be discursively and cooperatively interpreted and translated, which enriches this knowledge source with alternative interpretations. Still, it remains a challenge to integrate the outcomes of observational learning into the overall project.

> There was a discussion about the fit of pilot projects, but they simply did not fit each other. The project's starting point was that partners did their own projects. (Partner RIVERS)

For observational learning to work, knowledge needs to be generalised. This happens in three steps (Bandura 1979):

(1) *A wide variety of situations is observed that have a rule or principle in common.* Pilot projects thus require a certain degree of variation, while being related. In SEWAGE, pilot projects were directly linked and they were highly comparable, which made the insights in one case of relevance to other cases. The project made extensive use of the potential of observational learning and cooperation partners particularly learned from the much-advanced pilot project of the Swiss partner. In case of less coherent pilot project portfolios such as in RIVERS and PARKS, partners found it harder to relate to other pilot projects. The broad range of pilot projects in terms of development stage, size and focus was experienced as a challenge hard to overcome (*see quotation above*). In PARKS, for example, only project partners with pilot projects that dealt with the 'regional park' concept could benefit from a close relationship with their pilot projects (see section 5.2.1).

  Besides the strategic fit of pilot projects, opportunity is required to experience each others' pilot projects. Study visits provide unique options for 'experimental learning' by allowing access to tacit knowledge. However, they usually take place only once at each location and do thus limit ongoing learning. An exception was the PARKS project, where 'transnational feedback sessions' allowed partners to visit pilot projects multiple times.

> I experienced the transnational feedback sessions to be of little help. It is of course interesting . . . to see other pilot projects, but the truly important questions related to the project were never answered in this framework. I saw a lot, and I learned a lot, but this did not advance our pilot projects. (Partner PARKS)

> Reflection did not work as well as I had expected. Everybody reflected individually. This did not work optimally; the exchange should have been much deeper. We learned that you need time during feedback sessions to familiarise with the journal and to reflect. (Partner PARKS)

(2) *The rule or principle is extracted from diverse experiences.* This requires time for reflection at the individual and group level to increase the quality of learning processes (Ayas and Zenuik 2001; Kolb 1984; Engeström 1999), but also means that projects go beyond accumulating 'best practice' lists. Transnational reflection and reflective practice are further discussed in the next section.

(3) *The rule or principle is utilised in new situations.* This requires abstraction and generalisation so that lessons are transferable (see section 5.3.6). Pilot projects that include a knowledge development function require experiences to be turned into codified knowledge.

According to Bandura (1979), the ability to perform these steps depends on previous experience, skills, and motivational factors that determine which aspects of learned responses are translated into action. Pilot projects increase a project's share of tacit knowledge through their practice-oriented approach, which is harder to share and transfer. However, observing others' experiences directly offers options for conveying tacit knowledge that is otherwise hard to articulate.

Knowledge development requires the aspect of novelty, but in the case studies some pilot projects were limited to implementing concepts that had been applied before and where relevant experience already existed, rather than making new experience. In those cases, their knowledge creation function was obviously very limited.

PHASE 2: Conclusions

TRANSNATIONAL EXCHANGE AND COMMUNICATION

Transnational exchange allows partners to access each other's knowledge bases ('exploitation'), but also to jointly advance these and to integrate new experiences into existing knowledge stocks ('exploration'). Intense exchanges allow project partners to base their work on a pool of various knowledge sources and are thus a relevant first step of any social learning process.

In INTERREG programmes, the fact that projects exchange experiences and knowledge is often taken for granted, although particularly the exchange of tacit knowledge can be challenging. The case studies show that transnational exchange can be of different intensity and focus, which again influences its potential outcome. Transnational exchange is a very dynamic process: as in SEWAGE, it can increase in concreteness and result-orientation with time, but if there is little incentive to cooperate, partners may remain in a loose and even incoherent exchange as in RIVERS and PARKS. The case studies confirm that transnational exchange requires a reason and trigger for making knowledge accessible, such as joint work. Moreover, effective exchange not only needs time, openness and inclusion but an awareness of the value of partners' knowledge stocks and appropriate structures and platforms.

Fruitful exchange can only take place if project partners are interested in opening up to their transnational counterparts and value other knowledge and experience as useful for both their own purposes and for achieving transnational project objectives. Exchange can be based on different platforms, including study visits, partner meetings and group work. An interesting finding was the impact the design of work packages had on the intensity of exchange: work package design that followed a process logic integrated the different project aspects, topics and partners subsequently so that their inter-linkages required exchange and cooperation. Similarly, the structuring of meetings was decisive for the intensity of exchange: when partner meetings discussed individual pilot projects subsequently, exchange was less intense; when they followed a cross-cutting logic, discussion points were of concern for most or even all partners and these were able to engage more intensively.

Communication is pivotal to all cooperation and lays the basis for joint results by a joint effort. It is at the heart of the people-to-people approach between the carriers of knowledge and particularly relevant in the case of tacit knowledge. The objective of communication can be to ensure the flow of knowledge and experiences, to create new knowledge and to find consensus. Although it is rather trivial, communication was easier in cases of concrete subjects and tasks than in the case of loose connections between partners. Moreover, the set-up of pilot projects, discussion culture and trust impacted on communication intensity.

The communication flow in transnational projects is likely to encounter linguistic, cultural and political obstacles. Although partners in PARKS and RIVERS highly appreciated the diversity of partners and their approaches, they did not manage to address this in a way that could lead to joint results. Communication management can help to overcome many of the challenges posed by transnational cooperation and to establish a strong project communication culture. Facilitation, for example, can contribute to thinking beyond individual pilot projects and link up the different experiences and findings to form joint recommendations.

Transnational exchange and communication are potentially influenced by:

- the organisations' general fit and previous experiences that affect partners' motivation to exchange (see section 5.2.1);
- cultural and language factors strongly impact on project communication;
- the awareness of 'knowledge roles' and their translation into objectives for exchange and transfer as well as the awareness of the value of other partners' knowledge (see section 5.2.1);
- the general identification of partners with the project and its planned results strongly influenced partners' willingness to engage in exchange processes in the case studies and to make use of potential exchange options;

(continued)

(continued)

- knowledge characteristics such as relatedness and complexity, but also tacitness and context-dependency can affect the focus and target orientation of project communication (see section 5.2.2);
- the project's objective, defined tasks, working methods and division of labour that provide platforms for exchange, institutionalise it and/or require partners to work intensively on joint actions (see section 5.2.3);
- the strategic fit of pilot projects increases partners' interest to engage in exchange (see section 5.2.3).

Transnational exchange and communication potentially influence:

- how well transfer options are noticed and realised (see section 5.3.3);
- options for transnational feedback by knowing about partners, their motives, background (see section 5.3.5);
- options for joint knowledge development and learning. Without exchange, there can be no joint initiative, and the combination and joint interpretation of individual knowledge require intense exchange (see section 5.3.6);
- ability to identify and discuss crosscutting issues that span pilot projects and work packages (see section 5.3.6);
- systematising and abstraction exercises (for example key characteristic of the 'cluster evaluation' approach, see section 4.3.4).

CREATING A COMMON KNOWLEDGE BASE

To make efficient use of partners' existing knowledge stocks is an important challenge to projects and first and foremost requires sufficient knowledge about these stocks. It is concerned with the use of existing knowledge stocks during the cooperation process. Knowledge in a transnational cooperation project is multidimensional: it includes existing and developing knowledge, but also knowledge of factual, explicit, tacit and experiential nature. How to deal with the different dimensions and types of knowledge has not yet been well conceptualised in INTERREG programmes and leads to challenges in terms of efficient knowledge use, the integration of the dynamic dimension of project knowledge and the development of joint knowledge from case-based experience.

Self-evidently, the knowledge sources that feed the cooperation process are highly influential, being both starting points and a continuous influx. The more knowledge partners feed into their project, the more can be exchanged and transferred and finally processed to produce joint results. A strategic choice of partners thus also calls for a strategic selection of the knowledge portfolio that informs the cooperation and knowledge development process. However, in the reality of transnational cooperation, partners' knowledge bases are not always the prime selection criterion for participation.

In practice, project partners have different ways of integrating and further developing their knowledge. Tools can facilitate the compilation of knowledge stocks but remain both underused by projects and challenging in terms of project workflow.

The quality and quantity of project knowledge input is potentially influenced by:

- partners' institutional background, their previous experiences and knowledge bases, but also by their awareness for these and the translation of knowledge into action (see section 5.2.1);
- knowledge characteristics affect how easy knowledge can be used in the project (for example complexity) and how much attention it is given by transnational partners (value, novelty, relatedness) (see section 5.2.2);
- project objectives, project tasks, methods, partner integration, and fit of pilot projects (see section 5.2.3);
- communication intensity and exchange structures determine how partners share knowledge and experience.

The quality and quantity of project knowledge input potentially influence:

- knowledge characteristics: a high degree of newly made experiences implies a high level of experiential, tacit and/or highly context-dependent knowledge (see section 5.2.2);
- quantity and quality of knowledge fed into the process impact on how much and what can be exchanged and transferred and finally processed towards joint results.

EXCHANGE OF EXISTING KNOWLEDGE AND EXPERIENCE

Knowledge transfer is part of the raison d'être of transnational INTERREG programmes, but as the case studies illustrate, it does not happen automatically. The scope of knowledge transfer was limited even in cases of lively exchange processes. This is supported by the conclusions of Lähteenmäki-Smith and Dubois (2006) from their research on INTERREG projects, particularly in the case of direct copying processes. The interpretation of information is not a neutral process, and knowledge transfer implies that the recipient side assesses the relevance and validity of the information for their own context. Although such an assessment took place in WOOD and SEWAGE, the cases of RIVERS and PARKS show how challenging it can be in the context of transnational cooperation.

Although similarly concerned with the transfer of 'best practice', transnational cooperation projects differ from situations described by policy transfer literature as they describe multilateral learning processes. Nevertheless, concepts of policy transfer offer valuable insights into the degrees of and conditions of transfer. The role of different political actors, but also the relevance of the stage of a policy or innovation cycle is applicable to the INTERREG context.

Many of the structural parameters discussed in section 5.2 impacted on the ease or difficulty of knowledge transfer. The transferability of knowledge was, for example, strongly dependent on the relevant knowledge type and its contextualisation. With respect to the characteristics of knowledge, transnational INTERREG projects are not well placed for transferability. A potentially high degree of tacitness, ambiguity and complexity of the topics seriously limits the transfer of knowledge between project partners. A high degree of experiential knowledge developing in parallel pilot projects poses another challenge in terms of transferability and requires reflection and generalisation. The nature of knowledge is responsible for the stickiness of knowledge. Especially causal ambiguity and a lack of know-how and know-why are responsible for creating knowledge stickiness. Transfer can also be impinged by a lack of definition of the overall project purpose and too high divergence between pilot projects.

The case studies confirm that transfer objects related to technologies (SEWAGE), operating rules and know-how (WOOD, SEWAGE) and concrete instruments (WOOD) are easier to transfer than ideas, programmes and institutions (for example approaches to public participation in both RIVERS and PARKS). Helpful to knowledge transfer seem to have been concreteness of the transfer object, standardisation efforts, joint planning and intense communication to access sufficient background information on transfer objects. Moreover, the organisational fit and the strategic fit of pilot projects played a role. A lack of these severely limited transferability in RIVERS. Natural science objects that can be abstracted and verified made transfer easier in SEWAGE compared to social science issues that are more context-dependent, as in PARKS.

Knowledge transfer faces the conflict that very concrete transfer objects increase transferability as it is easier to relate to them, but if coupled with context-dependency, concreteness and specificity can become too strong and render transfer extremely difficult. In the case studies, many options for potential knowledge transfer were not used due to the inability to recognise transfer options as well as insufficient attention paid to context. For successful knowledge transfer, transnational cooperation requires specific effort, the coordinated assessment of experiences and joint communicative processes of sense-making and de-contextualisation.

The effectiveness of knowledge transfer is potentially influenced by:

- a gradient in existing knowledge as a starting point for knowledge transfer, the awareness and will of partners to act as senders and receivers; the fit of partners (see section 5.2.1);
- types of transfer objects, knowledge characteristics, such as tacitness, complexity or context-dependence and ambiguity, concreteness, standardisation, validity. A high degree of experiential knowledge from pilot projects requires reflection and generalisation to be transferable (see section 5.2.2);
- the strategic fit of pilot projects increases chances for mutual interest and transfer;
- intense communication supports sufficient background information on the transfer object.

Knowledge transfer potentially influences:

- access to other partners' knowledge and experience;
- identification of cross-cutting issues (see section 5.3.6);
- new insights and learning options.

EXCHANGE ON PILOT PROJECTS

Pilot projects are a commonly used feature in transnational INTERREG projects. They are the response to the programme requirement for 'tangible' results on the ground, but also to project partners' needs to achieve direct benefits at local level. Today's projects are much more focused on implementation than those of the past, which, however, also implies that the exchange of experiences has shifted to the making of experiences. This means that projects are faced with a time deficit for processing and reflection of experience.

Being a major transnational knowledge source, pilot projects can make use of a project's transnational potential by supporting 'observational learning' and by delivering experiential knowledge from more than one partner that can contribute to the development of new joint knowledge by reflection. Project partners can thereby learn from a variety of pilot projects while implementing only one. Observational learning requires the generalisation of knowledge, which is again based on three conditions: (1) pilot project portfolios need to have a degree of variation, while being related; (2) reflection of the relationship between individual pilot projects is a precondition for extracting joint findings; and (3) joint findings are applied and implemented. Particularly the second condition requires substantial time efforts. Previous experience, skills and motivation play a significant role in the ability of projects to perform these steps. Pilot projects solely focusing on action and learning-by-doing are not likely to allow more than the accumulation of experience, cannot fully develop their potential and remain limited to local use. As the exchange related to pilot projects influences further knowledge processing, transnational INTERREG projects and programmes can gain significantly by highlighting the potential benefits, barriers, and preconditions of the transnational potential of pilot projects. Project strategies are usually not conducive to observational learning, which faces a time dilemma: as pilot projects are planned before the project start, new experience and knowledge acquired during the process are hard to integrate into the running implementation of pilot projects. How well projects manage to share experience from pilot projects therefore strongly depends on the project strategy and, as the case studies show, faces many obstacles.

The four case studies confirm that partners often plan and implement pilot projects individually, which limits knowledge development to the individual level. Moreover, some pilot projects only implemented concepts of little innovative character. Thus, they had a limited knowledge function and at best a function for dissemination (for example introducing standards already in place elsewhere) and local knowledge development.

The exchange on pilot projects is potentially influenced by:

- partners' previous experiences; awareness of 'knowledge roles'; motivation for cooperation (see section 5.2.1);
- a strong implementation-orientation and little knowledge-orientation, differences in development stages and a high degree of context-dependence in pilot projects limit options for joint knowledge development;
- exchange structures, joint visits, a strategy that avoids the time dilemma, tasks that make use of the variety of experience (project examples in Chapter 6) support 'reflective practice';
- pilot projects' strategic fit, their degree of transnationality and a balance of variation and relatedness is necessary for observational learning (see section 5.2.3);
- communication for accessing partner knowledge and developing new joint knowledge.

The exchange on pilot projects potentially influences:

- options for inter-linkages between pilot projects, transfer and feedback; systemising new experience and abstracting new joint knowledge (see section 5.3.5).

5.3.5 PHASE 3: Reflection and Transnational Feedback

Joint discussion and reflection advance 'knowledge articulation' (see section 4.3.4), make experiences and results accessible to the whole project and increase the quality of learning in projects (Prencipe and Tell 2001; Ayas and Zenuik 2001). Without joint reflection, cooperation benefits are likely to remain at the local level (Prencipe and Tell 2001). They are also necessary for the second step of 'observational learning', the extraction of the principle from diverse experiences. Reflection can be specifically targeted at the planning and implementation of pilot projects, as in the PARKS project, which organised specific transnational feedback sessions. In general, reflective practice was very differently pronounced in the case studies and was particularly weak in RIVERS, where exchange was limited and if at all very ad-hoc. The SEWAGE project made use of a Scientific Advisory Board that provided regular feedback on the project progress. The project included a work package dedicated to reflecting and evaluating the experience from pilot projects.

A transnational partnership with often highly specialised partners represents a substantial pool of knowledge and experience, which can be used for joint reflection on new experience. Transnational reflection can focus on content, but can also include process aspects. Reflective practice enhances double-loop learning, which in turn helps to create transferable learning outcomes (Raelin 2001). At individual level, reflection supports the development of new individual knowledge; at group level it potentially allows 'transnational generative learning' and cross-fertilisation. However, judging from both the case studies and the survey in Chapter 6, 'reflective practice' has not yet been a standard element in INTERREG cooperation.

A particular form of transnational reflection is the feedback on the implementation of pilot projects. This is an additional feature of exchange on pilot projects and adds more direct and active forms of transnational interaction to the more passive observation of pilot projects. Transnational feedback can take shape in the form of advice to or even joint planning of pilot projects. Transnational feedback allows challenging interpretation schemes and reaction modes, potentially entrenched ways of thinking, sense making, routines and solutions. Like observational learning, feedback constitutes a specific potential of cooperation projects but complements the rather passive observation with more active elements and options for double-loop learning. Interviewees pointed to the fact that the mental 'distance' between transnational partners makes it easier to accept critical comments because transnational partners are seldom in a direct relationship to each other and consequences for daily work life are less likely. The PARKS project made particular use of this potential for transnational knowledge development, but it was also used by other projects. PARKS included particular transnational feedback sessions, which focused on one pilot project at a time (see next section).

Observations from the case studies suggest that the exchange on pilot projects depends on their purpose: pilot projects that tested new knowledge (as in SEWAGE) provided information of interest to other partners and allowed for joint generalisations and results. Several examples suggest a link between a perceived need for reflection and the type of planned results, for example, transferable decision-making tools. However, it has to be taken into account that trying to make sense of a limited choice of (often little related) experiences carries the risk of contorted perception and over-generalisation of individual experiences (Bandura 1979). This can be avoided by relevant discussions at group level.

Transnational feedback has (1) a *knowledge access function* but can also have (2) a *knowledge development function* when partners jointly reflect on observations and cross-fertilisation processes happen. The four case studies illustrate that feedback can be provided by various *sources*: project partners, additional colleagues and regional partners, scientific or advisory committees or external experts can all provide feedback to the work project partners are involved in.

The PARKS project mainly made use of the knowledge access function of feedback. Project partners met to jointly advise individual pilot projects. This seemed to work better when pilot projects were still in their planning phases. The knowledge development function of feedback was used in SEWAGE and led to partners mutually challenging interpretation schemes and reaction modes. Like other exchange processes, feedback processes can take place on an ad-hoc and more coincidental basis or in a facilitated or even institutionalised way.

Summarised, the types of transnational feedback found in the four cases included:

- ad-hoc feedback during study visits and presentations (RIVERS, access function);
- use of more experienced partner for feedback and advice (SEWAGE, access function);
- joint comparisons and analyses (SEWAGE, development function);

- institutionalised feedback in transnational feedback sessions that allow access to outsiders' views on their own region (PARKS, access function, partly development function);
- feedback processes required for joint products (WOOD, development function).

A Ad-hoc Feedback

> Feedback happened rather by coincidence, mostly when you visited a local pilot project. (Lead Partner RIVERS)

To a certain degree, RIVERS used ad-hoc transnational feedback either during informal time at project meetings or study visits (quotation above). This was usually limited to bilateral exchange and focused on crosscutting issues, explanations and alternative solutions to problems. Still, interviewees did not think that this potential of transnational cooperation had been made much use of and that partners had shared experiences only to a very limited extent. Partners did not have many opportunities to learn about each other's situation and framework conditions and time for reflection was not provided. It was thus not possible for project partners to truly understand and draw conclusions from other pilot projects. In this case, it may be of little surprise that this project that was challenged with highly different knowledge characteristics, pilot projects of little comparability and other inhibiting factors such as unequal participation, staff changes, a lack of interest in each other's activities, and a lack of recognition of and openness to the potential of transnational cooperation.

In WOOD, feedback came rather easily as partners cooperated on joint measures, which involved continuous discussions and reflection. The case of SEWAGE shows that partners' different development phases offer a potential for learning by feedback if a certain level of comparability is ensured. The fact that the Swiss pilot project in SEWAGE was finalised and provided results before other pilot projects even started, allowed other partners to draw directly on the Swiss experiences. Considerable and highly targeted feedback was provided in both directions – to the Swiss partner during the execution of their tests and from the Swiss partner during the implementation of other partners' tests. Supporting factors were the comparability of pilot projects, a strong discussion culture and the requirement for the coordinated activities and consensus evoked by the project's aim for comparable strategies and joint results. In the context of INTERREG cooperation, however, it needs to be taken into account that this example represents a rather uncommon situation due its strong research and natural science orientation.

B Institutionalised Feedback

Two of the case studies used more institutionalised forms of transnational feedback. SEWAGE made use of an 'Advisory Board' that provided feedback both to pilot projects and – more importantly – to the project's overall knowledge development process. Some of the project's major conclusions were developed during discussions that followed feedback from the Advisory Board and in cooperation with it.

> It was only an indirect exchange. Each region sent experts to qualify the different regional projects in concept and implementation. . . . That means that other pilot projects did not play a role in advancing innovation and exchanging experience. (Partner PARKS)

PARKS provides considerable insights into the preconditions and functioning of transnational feedback as transnational feedback was highly institutionalised and enforced in specific meeting formats for individual pilot projects. For each pilot project, this meeting format brought together a selection of transnational partners, who observed and reflected on the pilot project and provided advice. These meetings were mainly characterised by presentations on the progress of the relevant pilot project and partners trying to understand both the project and its framework conditions. The latter was particularly important as the project was highly influenced by diverging social frameworks. Feedback sessions included a limited number of project partners and some additional regional partners. They were supposed to ensure that 'partners [were] directly involved as advisers and consultants in the policy formulation and strategy development of other regions, sharing ideas, knowledge and experience' (PARKS project application). The objective of these meetings was to discuss learning processes across transnational borders 'so that regional partners and practitioners are able to apply that learning to the processes of planning, design and implementation of [their pilot projects]' and 'creatively modify the way they plan, design and implement projects' (PARKS Toolkit for Transnational Feedback Sessions). Direct advice was given to pilot project managers immediately after study visits and participants later noted their impressions in a journal 'to practice at least part of what is involved in learning from experience and at the same time render tangible evidence as to whether something has actually been learned or not' (ibid.).

> There was a Dutch corporate consultant in the transnational feedback session, who could not understand why we had a problem. He advised us to communicate to our colleagues that – instead of being a problem – we are part of the solution and to present ourselves in a more confident way. That was key to my colleague, and thus the transnational feedback session really made a change for her. (Partner PARKS)

> In the transnational feedback session transnational partners advised us to involve municipalities and citizens much earlier in the process. In [place name], federal planning processes then – for the first time – involved citizens in a brainstorming session before making a plan. Planning teams waited to start their work until after the participation. . . . These transnational interventions and impulses are not possible in decision-making processes but in planning processes. (Partner PARKS)

Feedback sessions worked rather *one-sidedly*, with participants reflecting on the development of one single pilot project at a time, meaning that feedback was case-based. This limited knowledge transfer options and challenged the abstraction and generalisation of transnational conclusions. The following reflection was mainly left to writing individual journals, which did not follow a common approach to content or methodology, and focused on content rather than on processes. Due to the particularly complex framework conditions in PARKS' pilot projects, transnational feedback sessions had to *evolve over time*. One interviewee identified two phases of these sessions: only after participants got familiar with the respective region and the particular project, could the commentary and consultancy function of the meetings shape up. The diversity and complexity of framework conditions inhibited the usefulness of feedback in many cases. Similarly to the overall project, feedback session lacked focus. Feedback was mostly anecdotal; only in very few instances were project-spanning issues and questions discussed.

In terms of benefit, one German partner reported that transnational feedback had triggered a reconsideration of roles in regional development (*see quotation above*). Other interviewees had more difficulties in pointing to more direct benefits from transnational feedback rather than a general impression of it as interesting to get different views on their challenges.

> I cannot simply go to partners in the Netherlands or in England, where they have completely different planning systems and cultures, and tell them what to do. I really have a problem with that. (Partner PARKS)

> You can learn directly from pilot projects, from the experiences of project managers, about what could be done differently and which mistakes can be avoided. In other projects, I gained a lot from these options, but in PARKS I was not able to experience anything alike. (Partner PARKS)

> There was a discussion with transnational experts who came to visit us. They started a debate on principles, about the fact that [river] should be of bathing water quality. I did not want to discuss this because it is simply not feasible. However, they kept coming back to this. I find that very difficult to deal with. (Partner PARKS)

Especially the purpose and potential benefit of feedback, as well as possible types of feedback, remained unclear to some participants (*see quotation above*). Feedback was particularly generic or lacking in the case of weakly facilitated sessions. These sessions were also not linked up to each other. Approaches to problem solving were rarely discussed. Some 'premise reflection' (questioning of suppositions, Mezirow 1991) took place at individual, but not at group level. On some occasions, reflection in PARKS also included negative lessons, which provided a high learning potential for transnational partners.

> I feel a bit disappointed about the seemingly small role of transnational feedback sessions on strategic orientation . . . and the decisions taken on

that basis. The feedback session had no direct contact with those doing the 'real-life planning process' . . . So I doubt they could live up to the high expectations formulated in the application. (Partner PARKS)

I was wondering . . . if this transnational feedback session was needed . . . at this stage. Sometimes it seemed as if planners had already thought about most things suggested. (Partner PARKS)

Moreover, differences in the development stage of pilot projects posed a problem for the timeliness of feedback. Several interviewees pointed out that giving advice was not of much use when the implementation of pilot projects was of little flexibility. Thus, the primary function of transnational feedback lay in influencing informal decisions and providing impulses (*quotation above*). Overall, feedback sessions seem to have been imposed in a top-down manner and in many cases only superficially applied and transnational feedback appeared to compensate for a missing overall cooperation and partly appeared to be a means in itself.

PHASE 3: Conclusion

Transnational feedback allows challenging interpretation schemes and reaction modes, potentially entrenched ways of thinking, sense making, routines and solutions. Similarly to observational learning, feedback constitutes a specific potential of cooperation projects and is strongly related to group reflection processes. It complements rather passive observation with more active elements and options for double-loop learning. Transnational feedback has both a knowledge access and knowledge development function. Despite its potential to add additional knowledge and innovation to project activities, transnational feedback is not yet systematically used in INTERREG programmes.

The extensive experience of the PARKS project with transnational feedback provides valuable insights into the requirements for beneficial feedback. Joint reflection and feedback require planning and effort, possibly institutionalisation and sufficient time. Feedback needs to be targeted; as a means in itself, its benefit is limited. Despite its elaborate feedback approach, the project was not able to capitalise on its full potential. This was due to a variety of factors, including a lack of facilitation and targeting of the concept, of time and of focus for reflection, pilot projects of very high context-dependency and complexity and thus a limitation to case-based feedback, as well as a lack of synergies and links between the different feedback sessions. On the contrary, in SEWAGE, the strategic fit of pilot projects, a strong discussion culture and focus on joint results ensured that transnational feedback was highly beneficial for the project progress.

To conclude, group reflection and transnational feedback represent a particular potential for transnational learning that makes use of the diverse knowledge stocks and can both enhance pilot project performance and joint knowledge development.

Options for group reflection and transnational feedback strongly depend on knowledge characteristics and a variety of project strategy parameters such as objectives, tasks, methods and pilot project fit.

The effectiveness of group reflection and transnational feedback is potentially influenced by:

- partner-specific aspects: relevant expertise in the field, motivation and openness, awareness of 'knowledge roles' (see section 5.2.1);
- knowledge-specific aspects: degree of context-dependency and complexity of pilot projects, tacit knowledge (see section 5.2.2);
- strategy-specific aspects: the flexibility, focus and strategic fit of pilot projects; differences in development stages, tasks and methods (appropriate platforms), meeting formats (facilitation, time to observe and discuss), clarity of project objectives, partner integration and division of labour, focus on joint results (see section 5.2.3A);
- communication is relevant for accessing partner knowledge, joint reflection and knowledge development.

Group reflection and transnational feedback potentially influence:

- reflection potential, new knowledge from experience;
- access to partner knowledge;
- focus and usefulness of cooperation, especially on pilot projects;
- joint knowledge development (finding crosscutting issues, abstraction) (see next section).

5.3.6 *PHASE 4: Approaches to Transnational Knowledge Processing*

By the end of their lifespan, transnational projects face the challenge of processing the knowledge that has accumulated during the cooperation process. Ideally, individual knowledge gains are combined and transnational knowledge gains identified and codified for dissemination. Joint knowledge is then developed based on reflection, interpretation and generalisation. This final phase in the knowledge development process is related to the concept of 'reciprocal learning capacity' (Lubtakin et al. 2001). However, the case studies show that the final phase is a considerable challenge for transnational projects and easy to neglect due to time constraints at the end of the project or a lack of interest in common results. They also show that one of the main challenges is to produce knowledge that all partners identify with, and that is transferable to other settings. Particularly the comparably short project durations and partners' fixation on pilot projects worked as barriers to this process.

Knowledge processing faces at least three challenges:

- The *first challenge* is to bring together individual knowledge gains and experiences from pilot projects and to systematise these (often scattered) insights into the various fields of interest (*storing, collating and systematising*). Otherwise, new knowledge remains at the individual (or regional) level.
- The *second challenge* is to go beyond individual insights and is connected to the identification of cross-cutting topics that ease collective reflection of experiential knowledge and potential common aspects (*finding common ground*). Partners can reflect at individual level at any time, but common knowledge processing lifts reflection to the group level to allow for cross-fertilisation and accessibility to all relevant information.
- The *third challenge* is to make sense of the diverse knowledge gains to codify new knowledge (*abstraction and generalisation*). This may be linked to very different evidence bases (pilot projects) and is often based on experiential learning.

These three steps build on each other; it is, for example, hardly possible to sort and 'classify' knowledge that has not been captured in one way or the other in the first place. Abstraction from experiential knowledge requires a certain overview of the existing insights.

Of the four cases, SEWAGE was least challenged with the challenge to process common knowledge, as the latter was operationalised in the logic of the work packages (analysis, tests, evaluation) and could build on highly comparable pilot projects. Particularly the comparability of experience is relatively atypical in the context of transnational INTERREG projects. SEWAGE also included a dedicated work package for assessing the various findings from pilot projects. In the other three case studies, joint knowledge processing proved to be much more challenging. Although RIVERS originally had a quite elaborate scheme for deriving transnational project findings from individual pilot projects, this was never actually accepted by project partners and abandoned after the project coordinator left midway.

The following section discusses how project partners in the case studies pursued their primary knowledge objectives and how this contributed to the consolidation of the different knowledge inputs and particularly to the development of new knowledge. It starts with looking at how projects collated and systematised knowledge as a basis for knowledge generation, if and how they were able to identify crosscutting issues that would link up the diverse pilot project portfolios and finally if and how they managed to make sense of individual insights to form new transnational knowledge.

A Collating and Systemising Knowledge

Given their varied knowledge sources, transnational cooperation projects face the complex management task of knowledge systematisation. This challenge is related to *project internal knowledge access, processing and usage* that are related to the 'combination mode' of SECI model (see section 4.2.2). By sorting, adding and categorising knowledge, partners gain an overview of the various knowledge inputs that have been included and knowledge outputs that have been produced during the process. New insights are captured and a basis is created for the reflection of individual experience.

In the case studies, several supporting factors for the systematisation of new experiences and knowledge could be identified. These include:

- *A clear project focus and reduced complexity*: The examples of RIVERS and PARKS illustrate how a high degree of complexity coupled with a rather unclear focus made the systematisation of the various new experiences extremely challenging. The more focused and less complex SEWAGE project faced fewer difficulties in keeping an overview of new insights. Managing and possibly reducing complexity can support knowledge systematisation, and if this is not possible working with conceptual categories may help common knowledge processing, as will be seen later.
- *Systematisation as a dedicated and ongoing project task*: Systematic internal knowledge processing can be a defined project task. In RIVERS, a small working group was set up to bring together existing knowledge and to evaluate the experience of pilot projects. Unfortunately, this was discontinued and findings not used. Instead, after the project ended, another small group tried to collate pilot project experience to produce the final report. However, this report did not go much beyond a description of pilot projects, and project partners could not identify with the derived recommendations.
- *Work package logic*: As all actions built on each other in SEWAGE, knowledge was almost automatically systematised during discussions. The high amount of partly unrelated work packages in RIVERS and WOOD meant that a large share of very different knowledge was produced that was not set in relation.
- Targeted discussion culture:

  > What each partner does is one thing, but we only need to discuss what we do together. It is nice to learn from each other but is this reason enough for a big data collection? (Partner SEWAGE)

  The management of the SEWAGE project ensured that all partners supported new knowledge and that a red thread connected all process parts. Discussions were highly focused and discontinued when not directly contributing to the main project objectives. Regular recapitulations of conclusions helped to record relevant aspects and to determine the next steps in SEWAGE. Work packages, thematic symposia and workshops contributed to structure discussions and exchange thematically.
- *Written interim products*: When systemising the experiences made in pilot projects, PARKS stands out with its journals in which participants reflected on learning experiences from other pilot projects. Partners also composed

reports on the experiences from own pilot projects. In addition, detailed minutes and reports from meetings and conferences served the same purpose.

- *Streamlined assessment of pilot project experience*: blueprints for the evaluation of different pilot projects were used in SEWAGE and allowed the accumulation of highly systematic and comparable knowledge. The additional survey (Chapter 6) shows that a considerable amount of INTERREG IVB projects worked with the systematic evaluation of pilot projects.
- Linking experiences from pilot projects to overall project:

> At some point, I would have liked an overview of the advantages and disadvantages of our different approaches. We could have compared what type of solutions we recommend for what kind of problem and what experiences we made with them. (Partner PARKS)

This is a particular challenge in the case of diverse pilot projects of little comparability or complementarity. In PARKS, project partners suggested the use of a matrix that would match the different experiences made in pilot projects with appropriate conclusions. However, the Lead Partner did not follow this up.

- *Project-external communication*: newsletters and web pages helped to structure and store experience and to synthesise and reflect new insights in WOOD.
- *Programme-internal reporting*:

> The dreaded report writing was decisive. It disciplines you. . . . Once a year, reports are written, and you can either see this as a formality, or you can write something useful. (Partner PARKS)

INTERREG programmes require projects to deliver regular progress reports, which potentially support knowledge compilation. To this purpose, they need to reflect achievements rather than simply list outputs and activities. Although not the case at the time of the WOOD, RIVERS and PARKS projects, the INTERREG IVB programme for Northwest Europe required projects to include a critical reflection on project achievements and potential problems. Accordingly, while progress reports in RIVERS remained descriptive and failed to reflect partners' experiences and lessons, reports in SEWAGE include reflections on partners' experience during pilot project operation. Also, interviewees in WOOD and PARKS found the summary of project activities in the progress report to be beneficial for the processing and structuring of relevant knowledge gains.

B Finding Common Ground: Working with Thematic Clusters, Conceptual Categories and Cross-cutting Issues

> You can reduce complexity by making use of clusters of topics. (Partner PARKS)

Having accompanied a transnational project during the production of its final results, De Jong and Edelenbos (2007) suggest that the existence and use of cross-cutting issues and overarching ideas and concepts plays a role in developing joint knowledge. These issues and concepts built a bridge between the overall project topic and the highly specific subjects of individual pilot projects. Similarly, the concept of cluster evaluation (see section 4.3.4) emphasises the role of thematic and conceptual categories for the formulation of common findings and overcoming the limitations of case-based experiences. Identifying conceptual categories can help to find crosscutting links between project partners and their pilot projects. They work as a tool for reflection on individual action in a larger context and as a valuable preparation for knowledge abstraction. Systematisation exercises of knowledge gains (see section above) support the cross-linking of individual experiential knowledge.

> The [sub-]topics were highly interesting, but they could have been more narrowed down and focused the process of formulating the project strategy in order to avoid this diversity of regional pilot projects. (Partner PARKS)

In terms of the evolution of cross-cutting topics, the case studies differed considerably. Both in WOOD and SEWAGE, common conceptual categories between pilot projects assisted knowledge development. Although some conceptual categories could be found in RIVERS and PARKS, these were not always shared and used by all partners.

- *Cross-cutting issues are given by work package logic*: Work packages were constructed according to the logical working steps that were necessary to deduce the project's conclusions and that all pilot projects ran through in SEWAGE. Consequently, these working steps were of relevance for all project partners.
- *Project topic-inherent crosscutting issues*: During the cooperation process in WOOD and RIVERS, project partners identified issues of strong interest to the overall project, but that had not yet been addressed by the work package logic or other actions. In WOOD, partners realised that working with sustainable forest management required a debate on forest biodiversity and set up a specific working group to collate relevant knowledge and determine the project's take on biodiversity. These issues helped to advance the general project subject and resulted in a common vision. In RIVERS, partners found that the different cultures of public participation as well as relevant EU legislation concerned all partners and joint sense making. Although relevant EU legislation was discussed among project partners, the project did not find a way to address different participation cultures and to reflect on their impact on pilot project results.
- *Pilot project-inherent cross-cutting issues*: In PARKS, some crosscutting issues emerged in communicative processes when project partners discovered that certain pilot projects had issues in common. This was particularly the case with visualisation techniques for green spaces. These issues helped to advance single pilot projects but were not used for findings at project level.

- *Pre-determined crosscutting issues*: In PARKS, several conceptual categories were pre-determined by the Lead Partner and 'institutionalised' in regular conferences (such as 'identity' or 'regional governance'). These were of a character, which some partners found too generic to relate to and they were subsequently not always taken up by pilot projects. As De Jong and Edelenbos stress, it is important to find the right balance between being 'specific enough to lead to a mode of thought, but not specific enough to preclude a variety of different interpretations for different people and contexts' (2007: 702).

C Abstraction and Generalisation

Transnational projects need to process their diverse knowledge inputs for further use beyond the immediate project. In this respect, case-based knowledge requires abstraction and generalisation to produce transnational project results. Codified and abstracted knowledge is easier shared with colleagues and allows organisations to be less dependent on the tacit and experiential knowledge of individual members of staff. The formulation of codified knowledge also helps to overcome 'learning boundaries' towards partners' home organisations (see section 4.3.4).

Cross-linking experience and knowledge domains requires sufficient time not only for discussion and reflection but also for conceptualisation and abstraction processes that produce common knowledge (Engeström 1999; Nonaka and Takeuchi 1995; Kolb 1984). Different perspectives need to be combined, and lessons learned recorded to integrate new knowledge. This corresponds to the 'externalisation' step of the SECI model (see section 4.2.2). Due to the evidence base for generalisation being built on a limited amount of pilot projects and experiential knowledge in transnational cooperation, attention needs to be paid to not over-generalising conclusions (Bandura 1979). The case studies show that further phases of knowledge creation (for example combination, interpretation, abstract conceptualisation) do not automatically take place.

As discussed in section 4.3.4, knowledge production of higher validity requires the generalisation of case-based knowledge bound to specific conditions and the 'de-contextualisation' of knowledge for external use (Potter 2004). This generalised knowledge then again needs to be applied to specific contexts or re-contextualised (Hassink and Lagendijk 2001). Generalisation is of particular importance when pilot projects produce context-dependent experience. In some cases, however, the design, conditions and thus outcomes of pilot projects can be too context-specific and the specific experience not representative and non-generalisable (Vreughdenhil et al. 2010).

In general, two approaches are possible: either projects design comparable or complementary pilot projects as in the SEWAGE project or embrace and deal with the challenges of diverse approaches. The first approach may not always be in line with the programmes' policy of funding concrete investments and visible outcomes rather than particularly scientific research designs. As the case studies highlight, joint interpretation of experiences and results does not always come naturally to transnational projects as this calls for a switch from action-oriented working to reflection. Specific learning tools at individual and group level, as well as the evaluation of pilot projects and their contribution to the overall project, can help to overcome these challenges.

An approach that deals with the inevitable diversity of approaches in individual projects is the concept of 'cluster evaluation', developed for programmes that consist of relatively autonomous projects but follow the same general objective while implementing different measures (Sanders 1997). This tool creates clusters with respect to similar objectives, strategies or target groups that allow deducting generalised insights for practitioners and decision-makers. The key methodological components of cluster evaluation include thinking in thematic categories, communication and dialogue, facilitation, negotiation and validation as well as participation (see section 4.3.4). These components form the basis for the assessment of the case studies' approaches to process and abstract knowledge.

The following section discusses the case studies' attempts to reflect experience and generalise knowledge. Three aspects were identified that played a particular role in knowledge processing: (1) the use of project evaluations to assess the performance of pilot projects, deduce their contribution to the overall project objective and thus to the transnational dimension of new knowledge; (2) the role that both the involvement of the relevant target group(s) and the result-orientation play; and (3) the ability to combine fragmented knowledge with the help of formulating final products and reports.

Evaluating the Experience

The challenge of de-contextualising case-based knowledge is mainly one of evaluating processes and products. Project evaluations support knowledge processing as they systematically investigate the contribution of individual actions to the overall project and the general development and result production, depending on their focus. Thereby, they enhance reflective practice in a project. As they stimulate reflexivity and structure collective learning experiences, evaluation activities can be used as an integral part of the learning process (Colomb 2007). In this context, evaluation has two dimensions (Chelimsky 1997): (1) it helps project partners to review their performances and achievements and to identify potential improvements (internal dimension), and (2) it supports the creation of new knowledge transferable to other contexts (external dimension). In projects with heterogeneous pilot projects, it is impossible to merely aggregate findings, and cumulating results is more complex than replication, aggregation and verification (Sanders 1997; Potter 2004).

In PARKS, RIVERS and SEWAGE, pilot projects were evaluated, although these were of a low strategic fit in PARKS and RIVERS. PARKS conducted three external evaluations, which were limited to the learning outcomes of individual partners, did not attempt to synthesise these and were not fed back to project partners. Their use was limited to informing the final report.

> Each partner presented their pilot project and [the evaluator] analysed in how far this was public participation or only public information. That was when we thought of changing the focus of our pilot project to include true participation aspects. (Partner RIVERS)

In SEWAGE, the evaluation was supposed to contribute a more integrated perspective on questions that had mainly been dealt with from a purely technical point of view. In RIVERS, the evaluation aimed at assessing the relevance and findings of pilot projects, to foster exchange between them and to produce a joint guidebook. Partners' participatory approaches were analysed as well as links between the various pilot projects. It was found that some of the pilot projects included too few participatory aspects, which led a few pilot projects to change focus. Overall, however, the evaluation was only discussed in a small circle during the project, and project partners did not accept the evaluation findings. In neither RIVERS nor PARKS did the evaluations substantially impact on the projects' ability to produce joint results. Reasons for this include a lack of 'communication and dialogue' as well as 'negotiation and facilitation', which – according to the concept of cluster evaluation – are relevant components of joint sense making in case of a diverse project portfolio.

Focusing on Results and Target Groups

> I would say that we did not have an approach to knowledge development and result achievement at the level of the overall project. (Partner PARKS)

Abstraction and generalisation of case-based knowledge are challenging processes that involve the synthesis and interpretation of various knowledge sources. Clearly, they are not a means for themselves, but projects are more likely to engage in these processes if they have a specific use for common findings.

> We worked in parallel, and it was not about achieving a defined final result. (Partner PARKS)

Truly *transnational results* were only produced in WOOD in the form of a variety of specialised brochures. As SEWAGE was a highly explorative project, the results of the joint evaluation showed that no clear recommendation could be given on the subject matter as different analyses (for example ecotoxicological vs. economic) came to different results. Both RIVERS and PARKS had planned to produce transnational products (guidebooks with policy recommendations) but abandoned these during the process. Their final reports are dominated by descriptions of the achievements of the various pilot projects and include only a few policy recommendations or 'success factors', which were developed solely by the respective report author. As these two examples show, knowledge generation is likely to be impeded when different components are not linked, that is when projected project results are not based on the project's overall knowledge gains.

You have to agree on the target group for a report and then approach it accordingly. If we had wanted to target the EU, we would have approached the whole thing differently. In [predecessor project], we achieved more knowledge gains, especially when I look at the two different final reports. . . . PARKS was more about demonstration. (Partner PARKS)

Both WOOD and SEWAGE *defined target groups* for their project results and engaged in discussions on how project findings would best be communicated. WOOD directly involved its target groups in various meeting formats. Although knowledge generation ran comparably smoothly in SEWAGE, project partners found the process of making sense of partly contradicting project findings and deriving policy recommendations highly challenging. This can possibly be linked to a lack of target group integration. Both RIVERS and PARKS were rather self-referential systems, where measures were implemented for partners' own purposes rather than for external target groups. Although PARKS' final report includes 'political statements', policy-makers were not involved or specifically targeted during the cooperation process.

Writing the Final Reports, Handbooks and Guidelines

I could not make any use of the final report. (Partner PARKS)

Transnational project results can be found in a variety of products such as guidebooks, reports and studies (see Chapter 6). In transnational INTERREG projects, an overview over these is usually provided in the final report, which can also include the project's conclusions and recommendations and thus potentially enhance reflective practice and project performance. The final report provides partners with a work of reference of their own, their partners and transnational lessons. Final written products can pursue different purposes and can be written for very different types of target groups. To represent the entirety of all lessons, they require appropriate discussion processes. Again, these can be of different communicative and inclusive character. For a real transnational representation, individual experiences need to be reflected and findings abstracted with which all partners identify. When project results include tacit knowledge, these are less likely to be found in a report. In these cases, direct and personal communication and user involvement can be of help (Vinke-de Kruijf et al. 2013; Koskinen et al. 2003).

One of our partners had to combine the bunch of pilot projects at the end. He took over the responsibility to write the final report because he had previous experience with tasks like that. This happened when the project was already finalised and got an extension to write up the final report. That made it very difficult. He surely tried as best as he could. It was more the fault of the overall situation and the partners that it did not go after plan. (Partner RIVERS)

The final report was written by the Lead Partner only. We were not really involved. . . . We sat with incredibly long texts and had to discuss these in a short time. He never accepted any of our remarks and wrote the report by himself. It was obvious that he did not want feedback. (Partner PARKS)

The question if we should go beyond addressing the political level and start working more intensely on the topic and process it against the background of theory and practice was a very difficult one. In this respect, we could not agree with the Lead Partner. We had very different points of view and the final report thus turned out to be a compromise. (Partner PARKS)

Table 5.5 provides an overview of the production processes of the final written products in the four case studies.

The overview shows that the case studies had quite different starting points and intentions for producing and processing knowledge for their final written products. SEWAGE, for example, aimed at producing joint and transferable results. The project discussed the relevance of different pollution substances and ways to deal with them. Knowledge generalisation was based on a project design in which pilots were of high comparability and jointly evaluated. Thereby, the project was able to transcend case-based knowledge. The well-established discussion and working culture and a strong focus on methodological questions prepared project partners for a systematic abstraction of findings. Continuous capturing of new knowledge and learning effects in discussions and reports helped to prepare the final report. Nevertheless, the project was not able to derive final conclusions due to inconsistent findings of evaluations with different foci. WOOD's final report only focused on the meta-level of project cooperation, but the project also produced a variety of more specialised brochures and a small working group produced a guidebook.

PARKS discussed experiences with different participation approaches and regional concepts. Individual learning processes were captured in journals but did not find their way into the written project documentation. Several workshops allowed for intense discussions that were based on cross-cutting questions, but again, these were not taken up in the final project report. As in RIVERS, the final report in PARKS was limited to a list of pilot project achievements and a few generalised statements. Both in RIVERS and PARKS the knowledge processes lacked a discussion of the final objective, focus and target group of the report. Reflection and abstraction were characterised by a lack of communication and inclusion so that project partners did not always identify with the conclusions. These two cases show that final reports can be limited to a list of so-called 'best practice', where individual experiences remain non-reflected at project level. Their transferability for both internal and external use is questionable (see section 6.3).

Target groups for the final written products were only defined in PARKS and SEWAGE. The inter-linking and abstraction of individual experience was an inclusive process in WOOD but limited to the reflection process of one person only in RIVERS and PARKS. Both a lack of partner involvement and options for

Table 5.5 The writing process of the final written products

Questions	WOOD	RIVERS	PARKS	SEWAGE
Focus	Reflections on cooperation in the project	General information on subject, description of pilot projects, 'success factors'	General information on subject, description of pilot projects, political statements	Develop recommendations for the usefulness of a centralised or decentralised approach to waste water treatment
Objective and target group	Not defined	Not defined	Political lobbying	Political stakeholders, other scientists
Basis for decision	Outline agreed by partners	No joint decision-making, produced after project was finalised	Outline decided by Lead Partner, no joint decision-making	Proposal for outline by Lead Partner, joint decision-making
Involvement	All partners, but dominated by Lead Partner	Small editing group, partners only delivered information on pilot projects	Partners composed sections on their pilot projects, otherwise dominated by Lead Partner	Lead Partner, Steering Group
Responsibility	Lead Partner	One selected partner	Lead Partner	Lead Partner
Feedback processes	Yes	In theory, but not used in practice	No	Intense feedback in Steering Group, additional feedback from partners
Analysis of individual experience	Provided by individual partners and summarised by Lead Partner	Mainly by author, little feedback from partners	Provided by individual partners and summarised by Lead Partner	Joint process of interpretation
Abstraction	By synthesis	By identification of links and synergies between pilot projects	By synthesis and individual interpretations	With the help of a joint analytical framework

Source: by author

feedback strongly influenced identification with the report and its conclusions later on. This is confirmed by studies on interdisciplinary collaboration that emphasise the importance of joint authorship for project reports (Bruce et al. 2004).

PHASE 4: Conclusion

COLLATING AND SYSTEMISING KNOWLEDGE

With their variety of partners and their cognitive frames, transnational projects need to capture and link up knowledge gains to make use of these as a basis for common findings. As emphasised by models for knowledge creation in groups, this requires the reflection of achievements (see previous section). The experience from particularly diverse pilot project portfolios as in RIVERS and PARKS is challenging to synthesise and, therefore, calls for a systematised approach to assessment and interpretation. The case studies allowed identifying some supporting factors for knowledge systematisation. These include a clear project focus on guiding knowledge collation and systematisation, knowledge collation and systematisation as a particular project task, targeted and structured discussions, transnational assessment of pilot projects and a systematic use of interim written project documentation. If systematisation exercises are postponed to the writing of the final results as in RIVERS, it becomes disproportionately harder to recapture all relevant insights and conclusions. On the contrary, actions that systematically build on each other provided a strong methodic structure and increased knowledge systematisation and processing in SEWAGE.

The effectiveness of knowledge systematisation is potentially influenced by:

- partner-specific aspects, such as motivational factors (see section 5.2.1);
- strategy-specific aspects, such as objectives, focus, tasks, methods, pilot project fit (see section 5.2.3);
- communication.

Knowledge systematisation influences:

- joint knowledge development.

FINDING COMMON GROUND

Conceptual categories can help to find cross-cutting links between project partners and their pilot projects. They work as a tool for reflection on individual action in a larger context and – by creating 'common ground' – as a valuable preparation for knowledge abstraction and joint project results. Especially in projects with highly diverse pilot projects, possibly even with diverging objectives, finding common ground is an extremely challenging process that gains from using conceptual categories to span differences. Projects can include mechanisms that help the identification of crosscutting issues in a communicative way as shown by De Jong and Edelenbos (2007).

For several reasons, however, the suitability of the four case studies for studying the influence of conceptual categories on transnational knowledge processes is limited. The fact that RIVERS and PARKS did not either find or use crosscutting issues to produce joint knowledge may explain why these projects had difficulties generating transnational knowledge in the first place. In PARKS, the effectiveness of its pre-determined conceptual categories was limited due to a lack of participatory and communicative processes and due to not being applied to pilot projects. On the contrary, pilot project portfolios designed for comparability such as in SEWAGE do not need to strengthen their common ground with the help of crosscutting issues. Therefore, the more strategic a pilot project portfolio, the more cross-cutting links can be identified between individual pilot projects.

The effectiveness of conceptual categories is potentially influenced by:

- knowledge characteristics (see section 5.2.2);
- strategy-aspects: focus, work packages, pilot project fit (see section 5.2.3);
- communication, intensity of exchange, participatory development;
- feedback and transfer may help to identify 'common ground' (see sections 5.3.3); systematisation can contribute to the identification of conceptual categories.

Making use of conceptual categories influences:

- joint knowledge development.

ABSTRACTION AND GENERALISATION

To process reflection efforts and to make project results useful beyond the immediate project scope requires knowledge generalisation and codification. However, in line with empirical findings from research on project-based organisations, the case studies illustrate that often only little attention is paid to the processes that are related to producing long-term results. As partners tend to focus on immediate project tasks and less on opportunities for learning and disseminating project results, learning in projects can thus create strong barriers to the continuity of learning beyond project boundaries. Cross-linking experience and knowledge requires time and the evidence base of a limited selection of pilot projects poses extra challenges to knowledge abstraction and de-contextualisation. Particularly RIVERS and PARKS are good examples of how challenging the conditions for knowledge generation are in transnational projects and how irreconcilable task- and result-orientation can be.

In general, two approaches to knowledge abstraction are possible: projects design comparable or complementary pilot project portfolios or they actively deal with the challenges posed by highly diverse approaches with the help of specific learning tools and the systematic evaluation of pilot projects and the overall project. *Project evaluations* enhance projects' reflective practice and structure learning processes. As the case studies prove, evaluation activities require partner communication and integration.

A *focus on the production of transnational results* helped to link different project components and form joint project results. Early cooperation with defined *target groups* through feedback processes further increased the usefulness of results. Project products that collect the different lessons learned and combine and de-contextualise these to form transnational and transferable findings provide reason to transnational knowledge processing. However, as the case studies show, transnational knowledge generation requires the involvement of project partners in a communicative process that helps partners to draw lessons from their pilot projects to ensure partner identification with the results. Abstraction and generalisation also require appropriate tasks and methods. Abstraction and generalisation need systematisation; otherwise partners are – as in RIVERS – faced with a plethora of individual experiences. In this respect, the richness of reflection in PARKS' various workshops could have gained from bringing together, categorising and analysing the various thought and knowledge strands and could have formed a basis for abstraction. In a nutshell, projects need platforms for reflection and abstraction, negotiation and validation of joint results, but also awareness of the necessity for transferable results.

Processes of knowledge abstraction are potentially influenced by:

- partner-specific aspects: organisational motivation impacts on interest in producing common results (see section 5.2.1);
- knowledge characteristics: motivation for joint knowledge processing (for example value, novelty), difficulties during the process (for example high complexity or context-dependency of knowledge) (see section 5.2.2);
- strategy-related parameters: objective to produce common and transferable results, methods for individual pilot projects assessment, strategic fit of pilot projects, levels of overall transnationality, production of interim products to anticipate the challenges of knowledge processing during the process and ease time pressure at the end of the project, participation and involvement of partners (see section 5.2.3);
- communication between partners;
- quality of process phases: knowledge fed into cooperation process, effective exchange, systematisation, finding 'common ground' to prepare development of joint knowledge.

5.4 The Connection Between Project Structures, Processes and Knowledge Gains

Evaluative research on cooperation processes shows that process and outcome variables are linked to each other. This implies that good processes are more likely to produce good outcomes (Innes and Booher 1999). This section, therefore, links the findings on influential structural and procedural parameters with a brief discussion of the knowledge gains and learning outcomes of the four case studies. This is not meant to be a systematic evaluation of projects' knowledge gains. Instead, a reflection is provided on the form and locus of knowledge gains, the type of learning effects and their linkages to the overall project results and project implementation.

The four projects are looked at individually and the learning paths of the four projects traced in the following way: (1) their knowledge gains are briefly analysed; (2) project structures that worked in a supportive way in the learning process are summarised; (3) project structures that worked in a disruptive way in the learning process are summarised; (4) the main features of the project processes are recapitulated; and (5) the parameters that had the largest effects on the process of knowledge development and learning are abstracted. This is concluded by the identification of a handful of parameters that proved to be effective across the case studies.

GENERAL IMPRESSION OF PROJECT RESULTS

A general impression of how well the case studies were able to achieve their objectives helps to appraise their learning processes. All in all, it is remarkable that two projects were not able to achieve their planned results: neither the guidebook planned by the RIVERS project nor the 'toolkit' planned by the PARKS project were produced. Overall, with respect to the produced knowledge in the four case studies, a few aspects were particularly eye-catching.

> INTERREG projects matter to promote topics at regional level. In our region, we achieved very tangible results having initiated a whole new planning culture in terms of governance and public participation. . . . I am very satisfied at regional level. (Partner PARKS)

Knowledge gains limited to local level: Knowledge gains in RIVERS and PARKS largely remain at individual and local level. They were not abstracted or articulated and thus remain inaccessible to external audiences. However, even at the level of individual partners and their organisations, it often remained unclear how new knowledge was used. Moreover, in only very few cases were knowledge gains perceived as project results and it remains questionable how far pilot project experience was translated into knowledge.

Regional cooperation benefits: Projects including intra-regional cooperation could achieve regional cooperation benefits. In some cases, the effects were quite tangible, such as the introduction of a sustainability certificate (WOOD) or newly established regional cooperation structures (RIVERS). In other cases, regional results were of more intangible nature, such as general trust building for future projects (PARKS).

Lack of codification and dissemination: The case studies largely confirm that temporary project-based organisations are an appropriate organisational form for knowledge creation, but less so for the preservation of knowledge after the project ceases to exist ('knowledge sedimentation', see section 4.3.4). Particularly in RIVERS, the codification of knowledge beyond the scope of the project was perceived as 'distracting'. Partners focused on immediate project tasks and less on opportunities for learning and disseminating project results, a phenomenon widely acknowledged by literature on temporary project-based organisations (Grabher 2004b; Scarborough et al. 2004; Sydow et al. 2004; Ayas and Zenuik 2001; Keegan and Turner 2001).

Project results remain in a vacuum: In RIVERS and PARKS, project results were not linked to a concrete implementation strategy but remained in a vacuum. Project partners had difficulties pointing to relevant target groups, how these would be reached and how the produced recommendations and political statements would be implemented. Although PARKS identified political decision-makers as their target group, it had no targeted dissemination strategy and project partners did not disseminate joint results regionally.

> During the last ten years, public participation approaches have gained more acceptance in our institution. Public participation was smirked at for a long time and only accepted when linked to additional funds. However, in the end we were able to see an effect even in financial terms, which is critical for our engineers. . . . Then everyone thought that participation had led to something valuable. (Partner PARKS)

> PARKS truly changed our way of thinking. To me, that is the added value of these projects. (Partner PARKS)

> We were able to diversify our services and have become more active in the field. (Partner RIVERS)

'Soft' project effects dominate: In all case studies, knowledge integration processes took place, including changed perceptions (for example towards marketing, participation). Even when activities were discontinued, organisations could enjoy positive effects from project activities, such as in the case of a pilot project in PARKS that was able to build up trust among the general public through increased participation. This turned out to be of benefit to the organisation on later occasions. However, most of the lessons integrated into the organisational sphere referred to individual learning and not to joint project results in the strict sense. In some cases, increased intra-organisational cooperation was identified as an effect of transnational cooperation. Processes of institutionalisation were harder to identify beyond the certification scheme in WOODS and greater openness to

participatory approaches in PARKS. In general, project partners seem to have had difficulties in 'recognizing opportunities to transfer and in acting upon them' (Szulanski and Capetta 2008: 519). The 'stickiness of knowledge' at various levels led to situations where learning options were not realised during the project and not transferred back into the organisations, particularly due to 'learning boundaries' (see section 4.3.4) caused by the high autonomy of project learning from daily organisational practices.

Although the case studies provide many insights into projects' learning processes, some are examples of very limited learning processes. A conducive and conscious handling of the challenges linked to the production of joint results and learning in transnational cooperation was only found in the case of the SEWAGE project.

WOOD

Due to a strong implementation focus in WOOD, *knowledge gains* could mainly be identified in terms of technical and very practical knowledge of rather low complexity (for example PR, certification) and can be characterised as knowledge of the 'know-what' type and single-loop learning. This is mainly due to the project's limitation to 'experience accumulation' in terms of 'learning mechanisms'. Thematic learning effects are fragmented and spread across a variety of project results (handbooks, brochures). One work group also had more complex learning processes, when the combination of partner knowledge led to knowledge of the 'know-how' type. New knowledge was largely shared among partners and made accessible to external audiences. As groups of partners were responsible for products and documents, the partnership as a whole became the owner of learning processes and project results (*group learning*).

From the point of project structures, WOOD was equipped with *strong and supportive structures*. These include a transnational partner structure ('regional cooperation'), a transnational topic and transnational motivation for participation (access to markets, standardisation). A common knowledge base built on professional similarities supported a joint language between partners. Partners' thematic and methodical knowledge was of different depths, which made learning and transfer attractive to partners. Learning was also aided by the relative context-independence of relevant knowledge. Knowledge generation processes related to natural science topics and technical aspects proceeded without much difficulty and had to deal with few external factors. These were relatively easy to control, qualify or even quantify. Due to high proficiency, discussions were well informed and included the analysis of causal relationships. At the same time, the high share of very practical knowledge led to a high actionability of knowledge. With respect to the project strategy, the project objective was of transnational character and partner objectives corresponded to this. Sub-objectives substantiated the fairly open overall project objective. Presumably the most effective strengths of the project were the similarity of the challenges partners faced, their deep knowledge of the field, the novelty, value and actionability of relevant knowledge and the fact that the project only worked with common actions instead of pilot projects.

At the same time, WOOD was *challenged* with high institutional diversity and interdisciplinarity as well as an extremely uneven integration of project partners in its actions, which limited transnational exchange and learning. Moreover, awareness of potential 'knowledge roles' was missing among some partners, especially among potential 'knowledge senders'.

> The approaches were the same and it was therefore very easy to combine these documents. (Partner WOOD)

> It is important to avoid too general documents but to link them to practical experience. Finding this balance between general, overarching work and local work that is adapted to the context is challenging. (Partner Wood)

The *cooperation process* in WOOD was characterised by:

- the delayed production of a state-of-the-art report as systematic knowledge input, which overturned its purpose;
- generally intense exchange and communication, although this did not include all partners (either because these were less motivated to exchange or little integrated) and mainly took place in separate work groups;
- high transferability of knowledge due to its practical, partly standardised, character and a high share of context-independence;
- knowledge transfer was characterised by an exchange of easily articulated 'know-what' that is causally linked to relatively concrete and explicit cognitions and makes the project a 'vicarious learning alliance' (Lubatkin et al. 2001);
- a strong result-orientation towards the production of transnational products, although these largely remain unconnected as the project did not attempt to generalise knowledge from its broad range of actions;
- and a focus on compromise and partner identification with project results, although some conclusions were found to be too general and irreconcilable with local conditions.

Parameters that had a particular strong effect on the process of knowledge transfer and learning included geographical proximity between partners (which led to a feeling of 'closeness', supported a common language, common challenges and knowledge about each other, transferability and usefulness of joint products), the structure of work packages (their partitioning of smaller workgroups) that partly limited exchange, project methods that focused on joint actions and excluded pilot projects, a generally good – if unbalanced – partner integration, a strong orientation towards joint results and implementation, and the inclusion of relevant target groups.

PARKS

In PARKS, learning outcomes were more diverse: some articulated *knowledge gains* could be found, possibly enhanced by the abundance of written material that helped project partners reflect more in general. Thematic learning effects concentrated on public participation and regional planning and were both of the 'know-what' and 'know-how' type. In some cases, even 'systemic knowledge' could be detected, while new 'strategic knowledge' was hard to find. Numerous options for 'observational learning' mainly allowed for technical learning but also included some double-loop learning on public participation and spatial visioning processes. Then again, some partners only gained knowledge based on their individual pilot project experience or even could not identify any learning effects. New knowledge was accessible in evaluation reports and meeting minutes, but still little shared among project partners. Knowledge gains were distributed analogous to the focus of pilot projects. This meant that new knowledge on public participation could be found among all partners, while knowledge gains in relation to regional development and governance could only be found among some project partners. Again, this illustrates how the potential of transnational cooperation, which lies in sharing existing and emerging knowledge, could not be used to its full extent. Project results were general 'political statements' that were produced instead of the originally planned guidebook. These were not strongly anchored in the partnership and not followed up by an implementation and application strategy.

PARKS was equipped with several *strong and supportive structures*, but also challenged by a variety of obstacles to transnational knowledge transfer and development. Supportive structures include low institutional and professional diversity, high levels of novelty and value of knowledge due to partners' different development stages, and pilot projects with a high degree of innovation and the application of transnational working methods that eased the formulation of both tacit knowledge (transnational feedback groups) and explicit knowledge (thematic symposia).

On the other hand, PARKS was *challenged* by a series of factors including very different spatial levels (local, regional, national) and thus pilot projects of highly different scales, a lack of identification with potential 'knowledge roles' particularly on the side of 'knowledge receivers', a very strong focus on local matters and on implementing individual pilot projects, a lack of geographical or functional proximity between partners and different local framework conditions. Partners' reciprocal learning capacity was rather weak, possibly due to a lack of a unifying vision and strategic motivations. Relevant knowledge was characterised by high context-dependency and specificity, but low actionability, relatedness and validity. Knowledge processing was comparably complicated, as the relevant knowledge types faced an abundance of uncontrollable, non-quantifiable external factors (for example target group acceptance). Although setting project objectives is key to project management, PARKS had to deal with a lack of concrete project objectives and clarity on the project's transitional aspects. In addition, the translation of objectives into specific tasks and methods proved to be challenging. Partners partly disagreed on the main project focus and only little effort was put into formulating both common challenges and projected project results. The project did not identify transnational tasks and methods beyond transnational feedback sessions, which led to a lack of a transnational division of labour and tasks targeted at producing project results. Finally, the purpose and potential benefit of the applied transnational methodology was ill communicated and pilot projects were of low strategic fit.

The *cooperation process* in PARKS was characterised by:

- high mutual interest of partners in each other's experience and knowledge;
- intense exchange of ideas, concepts, instruments and background information;
- extensive knowledge input through thematic symposia (including external experts);
- many options for observational learning and reflective observation through transnational feedback sessions;
- impeded communication and observational learning due to large differences between partner regions and their pilot projects as well as due to a lack of project reflection;
- project partners who went a long way to avoid 'uninformed', 'incomplete' and 'inappropriate transfer' by intense exchange processes;
- low knowledge transferability due to its strong context-dependency, complexity and 'initiation stickiness' (Szulanski 1996);
- in some cases, more abstract know-how could be transferred, which – at least partly – made the project a 'knowledge absorption alliance' (Lubatkin et al. 2001);
- missed chances to use extensive transnational feedback for the production of transnational knowledge;
- a comprehensive basis for knowledge sharing, storing and systematisation through abundant written material, but a lack of processing of discussion outcomes and the analysis and combination of the individual lessons learned;
- a lack of knowledge abstraction of large amounts of individual experiences and a lack of further processing of discussion and reflection efforts into joint externalised knowledge;
- despite knowledge processing characterised by thinking in thematic categories and intensive communication during feedback sessions and symposia, the deduction of conclusions was insufficiently facilitated, negotiated and validated, and it remained non-participatory, with an unclear contribution of individual pilot projects to the conclusions.

The potential of PARKS' approach was undoubtedly high, including theoretical contributions, reference cases, focused workshops as well as the use of discussants and cross-cutting questions. To understand more complex 'know-how', partners tried to grasp related 'know-what' to understand causal relationships.

Such 'knowledge absorption alliances' involved intense exchange and learning processes, but were not used to their full extent. Not all experienced partners found cooperation partners at their level of experience and thus lacked inducement to question existing knowledge. While pilot projects allowed 'experience accumulation', formal learning mechanisms such as feedback sessions and project evaluations made 'knowledge articulation' and partly 'codification' possible both at individual and project level. The project thus corresponds to the 'navigator type' (see section 4.3.4), where knowledge articulation created an arena for double-loop learning, which, however, was little used at group level.

However, the achievements and lessons learned did not truly find their way into the project results. Symposia were a good starting point for the application of conceptual categories; they were participative and included processes of communication and dialogue. They had real potential for de-contextualisation through the use of crosscutting questions but did not link up and accumulate individual findings at project level. Transnational feedback sessions were communicative processes, but seldom included conceptual categories and processes of knowledge de-contextualisation. Some journals contained comparisons of pilot projects, but this remained unanalysed and non-accumulated at project level. Understanding partners' framework conditions was a first step to increasing knowledge transferability, but a systematic boost of transferability was not followed through. All in all, lessons learned thus largely remained at the individual level and were not further elaborated and generalised at the transnational level.

Parameters that had a particular strong disruptive effect on the process of knowledge transfer and learning include project structures that did not support exchange between pilot projects and that did not link up tacit knowledge gained in pilot projects and explicit knowledge acquired in thematic symposia. Moreover, the strong context-dependency of knowledge and unfavourable knowledge characteristics (for example low 'visibility', high ambiguity) as well as large differences between pilot projects led to high knowledge complexity and hampered exchange and observational learning. Institutional similarities and partly similar regional challenges supported transfer and learning, but a very ambiguous project objective and a lack of projected results coupled with strong local interests questioned the purpose and benefit of transnational working methods. A lack of communication with and integration of partners in the formulation of transnational results as well as a lack of result-oriented tasks limited the production of project results beyond the local level.

SEWAGE

In SEWAGE, *knowledge gains* were achieved in terms of new and deeper knowledge on the subject as such and in some cases also about methods (such as data measuring and interpretation). In general, discussions had larger chances to lead to 'strategic knowledge' than in the other projects, but the general impression was that partners mainly learned within their existing frames (single-loop learning) at a level that seemed beneficial for all.

From the point of project structures, SEWAGE was equipped with *strong and supportive structures*. These include the existence of deep expert knowledge that allowed highly targeted working in the project, a good exploitation of existing 'knowledge roles', identity of problems in partner regions and transnational motivation to participate (both exploitation and exploration). Partners were of strategic fit with a common knowledge basis and language, but differences in development stages. SEWAGE is an example of high reciprocal learning capacity, living up to most of the relevant preconditions identified by Lubatkin et al. (2001). Relevant knowledge was relatively context-independent and of relatively low complexity but high validity, comparability, actionability and relatedness and thus of high transnational value and transferability. Knowledge generation processes related to natural science topics and technical aspects and had to deal with only a few external factors. In case of external factors, these were relatively easy to control, qualify or even quantify. As the project dealt with verifiable facts rather than socially constructed phenomena, it was not as challenged with controversy as other projects. Method-wise, the project and its pilot projects were streamlined. SEWAGE had a very clear project objective, with an explicit transitional aspect, congruence of individual partner objectives and transnational knowledge as the projected project result. The project had strong 'task interdependence' (ibid.) and work packages followed a process logic, which supported partner integration, a transnational division of labour and result-orientation. All pilot projects were of a strong strategic fit and contributed directly to the overall project findings.

At the same time, SEWAGE was *challenged* with a lack of partners who possessed strategic implementation knowledge, a lack of integration of its primary target group(s) and a lack of a 'transnational issue'.

The *cooperation process* in SEWAGE was characterised by:

- intense and highly targeted communication and exchange;
- exchange and joint and systematic analysis supported by the overall project strategy with high partner integration, division of labour, and work packages following a process logic;
- strong support for observational learning through the comparability of pilot projects and their analytical and systematic dimension;
- good options for and use of knowledge transfer, which included abstract know-how and makes the project a 'knowledge absorption alliance' (ibid.);
- reflective observation through a well-developed discussion culture and an intense use of transnational feedback;
- high potential for knowledge abstraction through 'externalisation' and 'interpretation' as well as processes of deductive reasoning at group level.

The development of new knowledge was strongly based on comparisons: transnational cooperation made it possible to combine, analyse and compare different pilot projects. De-contextualising knowledge was less of an issue. Each work package produced a report that served as an information basis to the subsequent work package. One of the work packages was dedicated to the generalisation of overall conclusions. In terms of 'learning mechanisms', SEWAGE corresponds to the 'navigator type', where group work concentrated on 'knowledge articulation' that created an arena for double-loop learning. Of the four case studies, SEWAGE was the only project that managed to develop abstracted and codified knowledge that can, for example, be found in research publications.

Parameters that had a particular strong effect on the process of knowledge transfer and learning included a highly targeted knowledge input (ad-hoc partner feedback, institutionalised feedback by Advisory Board, state-of-the-art report) and knowledge output, a high motivation to learn from each other, the comparability of pilot projects and a joint analytical framework and work packages that followed a process logic. Moreover, the existence of joint and clear objectives and knowledge of low complexity and context-independency, but also a lack of target group integration were important factors.

RIVERS

In RIVERS, learning benefits at project level were difficult to identify. Instead, they were widely dispersed at individual level and dependent on the respective development stage and focus of the organisations. Lessons were not jointly processed in a reflective way. Some of the less experienced partners gained new knowledge of the 'know-what' type and deeper insights were acquired in the more innovative pilot projects. Partners with less experience in public participation did not seem to have had larger knowledge gains due to their limited commitment to the project's transnational purpose and reservations towards more innovative approaches. The project's strong focus on implementation led to knowledge gains of mainly technical and practical nature (for example participation methods) and thus related to single-loop learning. New knowledge was not made accessible to external audiences and the depiction of the implemented pilot projects in the final reports remained purely descriptive. In a way, RIVERS was the opposite case of WOOD and SEWAGE: all experience was made individually and even so-called 'common actions' were implemented individually (*individual learning*). Project results were general 'recommendations' produced instead of the originally planned toolkit. They were not strongly anchored in the partnership and were not followed up by an implementation and application strategy.

From the point of project structures, RIVERS was equipped with several *strong and supportive structures*, but also challenged by a variety of obstacles to transnational knowledge transfer and development. Supportive structures include low institutional and professional diversity that helped to develop a common language, geographical and functional proximity between partners ('axial partnership'), similar knowledge bases with differences in development stages, a 'transnational issue' and a transnational strategy for producing common results.

On the other hand, RIVERS was challenged by a series of factors including value diversity that developed from a potential into a barrier for knowledge transfer and unfavourable knowledge characteristics, including high context-dependency, complexity and some knowledge that was of novelty, but not of value to transnational partners (due to different traditions and cultures), while other knowledge was neither of novelty nor value to partners. A high share of practitioners led to a rather high share of tacit knowledge. The project did little to address knowledge complexity (for example by clarifying objectives). Participation was strongly characterised by local motives, but partners' interests in 'exploitation' remained underused. A considerable lack of 'absorptive capacity' (Lubatkin et al. 2001) can be identified, as partners were not always able to recognise the value of each others' experience and knowledge due to a lack of realised potential 'knowledge roles', particularly on the side of potential 'knowledge receivers'. This may be responsible for partners not being well prepared to understand potential transfer objects, but also due to a lack of a strategic choice of 'learning partners' (Pérez-Nordtvedt et al. 2008). In some cases, experienced partners even withdrew from active cooperation as their expected benefit from cooperation declined during the project. The project objective remained ambiguous and was poorly linked to the action plan; anticipated project results were not clearly identified. Project partners had little 'goal interdependence' (Lubatkin et al. 2001). Pilot projects were of low strategic fit, also because they related to different spatial levels and were thus of different scales. Partner integration was unequal and responsibilities for central tasks such as transnational knowledge processing remained unassigned until the very end.

> We had repeated discussions about how to combine pilot projects, but we never reached a conclusion. It was simply no one willing to take the reins and develop a concept. (Partner RIVERS)

The *cooperation process* in RIVERS was characterised by:

- the delayed production of a state-of-the-art report as systematic knowledge input, which largely overturned its purpose;
- low communication intensity and discussion of relevant partner differences led to limited exchange, which again limited options for observational learning;
- exchange structures that limited options for reflection and knowledge abstraction and focused on individual presentations on pilot projects;
- although interactions usually bring actors closer together, in this case, they contributed to some actors drifting apart and withdrawing from cooperation;
- a lack of use of observational learning options;
- limited transferability of knowledge and experience and failed transfer attempts;
- if at all, knowledge transfer was characterised by an exchange of easily articulated 'know-what' and made the project a 'vicarious learning alliance' (ibid.);
- knowledge transfer that was limited due to 'inappropriate transfer' attempts and both 'initiation' and 'implementation stickiness' caused by a lack of adoption of 'knowledge roles' and communication gaps between sources and recipients;
- staff changes;
- a lack of acceptance of the evaluation exercise and the general methodological framework that was supposed to synthesise pilot projects and which was subsequently abandoned;

- a lack of storing and sharing of lessons learned;
- a lack of cross-cutting issues and conceptual categories to link up pilot projects;
- a lack of knowledge abstraction, particularly reflection – the required step to turn experience into knowledge – was not much practised;
- the deduction of conclusions did not involve thinking in thematic categories and did not take place in a communicative, participatory, facilitated or negotiated way and unclear contribution of individual pilot projects.

In a nutshell, the RIVERS project was not focused on learning. Project part-ners recognised the potential of the project but were not sure how to 'harness the opportunities' (Sense 2003). A variety of structural parameters negatively impacted on partners' 'reciprocal learning capacity', such as a lack of organisa-tional fit, of a joint vision, strategic motivations and of learning intentions. The project emphasised 'experience accumulation' processes and lacked formal learn-ing mechanisms. It thus corresponds to the 'explorer learning type' (Prencipe and Tell 2001), where many activities took place at individual and informal level. This resulted in a limitation to single-loop learning processes.

Parameters that had a particularly strong effect on the process of knowledge transfer and learning include a lack of pilot project fit, a lack of focus of the project objective(s), a strong focus on local matters and a lack of instruments for exchange, discussion and systematisation of knowledge. These aspects are again related to others, including unfavourable knowledge characteristics such as low knowledge novelty and value, high context-dependency and unequal partner integration. The fact that the final project results did not require cooperation in a way that joint strategies or agreements require may have led to partners losing cooperation out of sight.

Knowledge Gains: Conclusion

The general *learning intensity* in the case studies depended on actual partner participation as well as the 'strategic fit' of pilot projects with the overall project and with the development stage of individual partners. Learning effects were not necessarily the largest when project partners focused on their local pro-jects. Lower learning intensity was found in cases where partners lacked similar organisational values and a unifying vision and strategic motivations, which are responsible for the reciprocal learning capacity in partnerships. The case studies show that more substantial learning (double-loop) took place in situations char-acterised by common, complementary or comparable activities, observational learning and transnational feedback. They also confirm findings from policy transfer literature that ascribe levels of higher transferability to aspects of the 'operational level', such as procedures and techniques (De Jong and Mamadouh 2002; OECD 2001).

The multiplicity of pilot projects allowed considerable 'observational learning' in the case studies. Although a substantial amount of project partners was inter-ested in transnational learning, many did not adopt a 'receiver' or 'sender' role for knowledge transfer. Beyond the transfer of existing knowledge in 'reciprocal learning alliances' (Lubatkin et al. 2001) by a constructive integration of different inputs, synergy effects also lie within the shared, inter-organisational knowledge base and allow learning *with* each other. This, however, requires partners to cooperate truly and engage in more than individual pilot projects, such as when producing a common product or process as shown in WOOD and even better in SEWAGE. These activities, then, also have a common knowledge-objective, which structures and guides learning. The type and strength of cooperation thus also depend on the type of planned project result(s).

In the case studies, a majority of lessons learned remain at individual level and were not made accessible to cooperation partners and beyond. It can be concluded that opportunities to walk unknown paths, to reflect on their own experiences, to discuss with transnational partners, to use options for observational learning and to engage in transnational feedback increased and deepened learning processes. To abstract lessons from others' experiences requires partners to take the time to understand cooperation partners and their contexts. This again can be supported by dialogue and cross-fertilisation processes.

Comparing the PARKS and RIVERS projects illustrates some interesting contrasts. In RIVERS, certain structural parameters laid a good foundation for learning and knowledge development processes. This was the case with partnership- and knowledge-related parameters, such as a spatially and functionally interrelated partnership and a transnational topic. Strategy-specific parameters were, on the contrary, a weaker starting basis, especially with respect to the coherence of objectives and actions, the concreteness of objectives, participation of all partners and the fit of pilot projects. The identified process phases were weakly developed. In PARKS, on the other hand, structural parameters laid a weaker basis for cooperation, including a lack of a transnational topic and a spatially or functionally inter-related partnership. Project-strategy related parameters were partly better developed, especially with respect to methods. Most importantly, the project managed to encompass several of the identified process components in a productive way (through means of symposia, feedback sessions, learning logs, regional reports, minutes) and was thus able to involve an intense discussion and reflection process. Here, the main barrier to transnational knowledge production was the lack of actual transnational knowledge processing and the generalisation of transferable knowledge. The two projects were characterised by a similar lack of discussion and reflection platforms to produce joint, abstract findings. Moreover, the writing processes for the final report were neither communicative nor participative and did not include negotiation and validation processes.

It should not be forgotten that only parts of the knowledge produced by projects are of explicit nature and can be codified. 'Handbooks', 'guidelines' and similar products can only ever include parts of the project findings. Moreover, as the examples of RIVERS and PARKS show, final reports and products are characterised by so-called 'good practice' descriptions that – particularly in RIVERS – suffered from a lack of reflection and evaluation.

The cross-case analysis of parameters with a particular strong effect on processes of knowledge development and learning points to six factors that played a role in more than one case. These include:

- the strategic fit of partners and pilot projects;
- partner motivation to participate;
- knowledge complexity;
- formulation of clear objectives and planned results;
- methods (work package logic, joint actions, methods for exchange, systematisation and abstraction, result-orientation);
- integration of project partners and target group(s).

6   Transnational Working Approaches
A Survey of Northwest Europe

Despite the rich insights the case studies provide, none of the cases was a satisfying example for applying transnational methodologies to tackle the main challenges of transnational knowledge development. SEWAGE, with its strong nature science focus and more or less context-independent knowledge, is a rather atypical case and of the three other projects only PARKS worked with specific working methods and tools for transnational cooperation. Still, these did not allow making full use of the project's transnational potential. A small survey conducted on a larger, and more recent, selection of INTERREG B projects provides further insights. During the IVB (2007–2013) funding period, programmes generally emphasised stronger transnational approaches.

Methodologically, scheduled actions are based on an understanding of how to achieve project objectives. In principle, a project strategy describes the epistemological concept and how objectives are turned into results. Moreover, an understanding is needed of how the project strategy is related to the project's transnational character; that is, how to make optimal use of the transnational partnership or how to overcome relevant barriers. At the centre of a transnational project strategy are the challenges posed by the duality of addressing a common issue, following common objectives and producing common results on the one hand and implementing individual pilot projects on the other. Work packages and actions then operationalise the general project strategy.

The survey analysed methods used to actively approach knowledge development in transnational cooperation in the INTERREG IVB funding period. Relevant data was harvested from the project applications of 50 approved projects in the Northwest Europe programme. For the in-depth project examples, interviews were conducted with Lead Partners. Projects in the NWE programme had to describe the transnational added value of each of their work packages and methods to this end. The survey, therefore, focuses on the logic and structure of project work packages.

The following section discusses the types of project methodologies found in the survey in terms of supporting the development of transnational exchange and knowledge. In this respect, two major groups of projects were identified, which are analysed with respect to their methodological elements. This is completed with an in-depth analysis of five transnational projects that used particularly elaborated methods for the exchange, transfer and development of knowledge.

It is not yet mandatory for INTERREG B projects to describe their general methodology or specific methods they will be applying or to consider potential causal relationships between project objectives and intended results. It is striking that several projects in the survey argued to have truly transnational objectives and transnational results but do not describe how this will be achieved and how individual pilot projects will contribute to this end.

In the dialectics of INTERREG B projects, many projects produce 'toolkits', 'manuals' and 'guidelines'. In the history of the programmes, these are attempts to overcome the often rather untargeted exchange of experiences and the limited European relevance of local and regional investments of the early programme days. These project results are still popular, although there is no general differentiation and conceptualisation of the terms at project or programme level, and they seem to be applied interchangeably. Other popular project results include position papers and management tools such as business plans, PR strategies, accreditation schemes, joint standards, and common labels. Only in very few cases do projects explain how these results will be achieved when working with case-based examples. It is surprising that only very few projects aim at producing a common vocabulary and joint definitions related to their topic. Taking the diversity of planning and societal systems in Europe into account, to agree on the terms to be used seems key for transnational cooperation (cf. Healey 1992). Still, projects may simply not consider this challenge at the application stage.

In the survey, all analysed projects included individual pilot projects and exchanged knowledge and experience between these in one or the other way. Looking closer at the 50 projects, two main groups can be identified (see Figure 6.1):

a   The logical starting point for transnational cooperation is the implementation of individual pilot projects (type A). Projects then use joint evaluation activities to follow up individual experience from pilot projects, while others work with transnational feedback that helps to advance pilot projects.

b   The logical starting point for transnational cooperation is a transnational element that precedes the implementation of individual pilot projects (type B). This is can be a phase of common knowledge gathering or of joint conceptual work. Some projects also include an additional downstream element, where measures are harmonised or activities coordinated as a result of cooperation.

6.1 Project Type A: Focus on Pilot Projects

From a methodological perspective, 15 of the 50 analysed projects worked with pilot projects as their logical starting point. These either contained an element of mutual feedback or a cross-pilot project evaluation; in two cases, pilot projects were jointly implemented. A few projects worked with both feedback and evaluation. In all cases, projects run the risk of postponing joint reflection and interpretation until a point where they find that 'it is difficult to compare cases that are hardly comparable' (partner RIVERS). Instead, the assumption seems

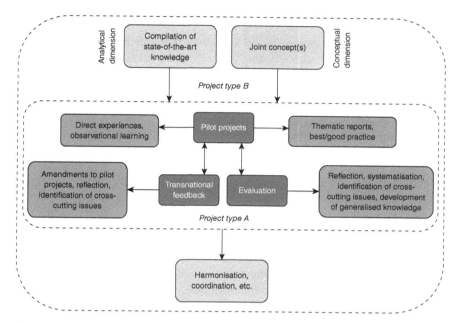

Figure 6.1 Methodological projects types in the INTERREG IVB Northwest Europe

Source: by author

to be that pilot projects in one way or the other automatically deliver transferable knowledge and expertise or autonomously identify the best applicable practices.

Transnational feedback allows reflecting on pilot project achievements and identifying potential crosscutting issues between them. Moreover, partners can also jointly plan, implement and improve pilot projects or even jointly coordinate related policies.

Examples of *evaluative elements* in transnational projects include the comparison of differences and similarities between pilot projects in order

1   to identify the best option of a prototype;
2   to assess the transferability of pilot projects;
3   to assess regional baseline data for a subsequent comparison with the situation after pilot project implementation; and
4   to make use of benchmarking exercises.

The basic idea behind all of these approaches is that partnerships can jointly conclude how to deal with the addressed problem based on the assessment of pilot projects.

6.2 Project Type B: A Common Starting Point of Transnational Relevance

From a methodological perspective, 35 projects based their cooperation on an element of transnational relevance that logically preceded the implementation of individual pilot projects. In 18 projects, this common element was of analytical character and focused on the creation of a common knowledge base with a review of state-of-the-art knowledge/existing best practice or the determination of regional framework conditions and baseline data. The implementation of pilot projects is then supposed to build upon the achievements of this initial *knowledge-gathering phase*. However, as seen in the case studies, in the reality of project life, this logical chain of action is not necessarily expressed in a chronological way but can run in parallel to pilot project implementation. The extent to which pilot projects then manage to take relevant findings into account depends on the nature of both pilot projects and findings as well as on the project's overall flexibility.

In the other 17 projects, the common phase involved creating a *conceptual basis* for the project, which was then implemented by pilot projects. Thereby, pilot projects carry common traits, which ease their contribution to the overall project objectives and the application of common assessment criteria. Examples of common concepts include the joint development of:

- techniques;
- priorities and a common vision as a basis for decision-making;
- information systems as a basis for investments;
- methods and indicators;
- methodologies for the assessment of pilot projects or for cooperation in general.

A common conceptual basis can be linked to a 'reality check' tool when pilot projects are used to assess the feasibility of a theoretically developed strategy. An illustrative example is the MANDIE project (see section 6.4). Finally, a few projects also included the harmonisation of measures.

A general impression from the survey is that although many projects claim to produce transferable results, they usually fail to explain *how* these will be achieved. On the contrary, many projects seem to understand the transferability of results as providing a list of good practice.

6.3 The Good Practice Concept

Many EU programmes support the production and use of best practices, including the INTERREG programmes. Although the role of examples and lessons has been widely recognised in planning and policy-making and despite its wide application, the term 'good practice' or 'best practice' (in the following, the two terms will be considered as interchangeable) often goes without a definition and explanation of why these practices are appropriate. The EU URBACT programme defined 'good practice' as an effective solution to a problem found by policy-makers and project managers or the demonstration that this solution is better than others (URBACT 2004). INTERREG programmes equate 'best practice' with learning from the partnership's experience (Northwest-Europe INTERREG IVB Programme 2010) or as an example for transferable results (Baltic Sea Region Programme 2012). Another definition from the context of European planning specifies 'good practice' as 'structured information . . . about successful experiences in local contexts, concerning issues generally acknowledged as relevant, evaluated according to a set of criteria' (Vettoretto 2009: 1069).

All the above definitions imply that the practice is recognised as appropriate, regardless of its specific context and situation and on a certain evaluation that asserts its appropriateness and of it being 'good' or 'best' in relation to others (ibid.). The word 'practice' implies routine and repetition and an occurrence that is never unitary or individual (ibid.). In light of the diversity of European member states with their considerable disparities in governance, administrative cultures, historical embeddedness, social and legal systems and economic situations, the transferability and validity of best practice, however, has to be questioned. Instead, the 'making of best practice' requires the validation by consensual definition of issues, solutions and evaluation criteria and a process of reducing complexity by strategic de-contextualisation from the political, social or historical context (ibid.; Bulkeley 2006).

In practice, the production of 'good practice guides' is essentially based on the belief that knowledge about specific projects is a helpful means for improving policy and practice in other countries and regions. Policy makers and practitioners on their constant search for 'ready-made, off-the-shelf policies . . . that can be quickly applied locally' (McCann and Ward 2010: 175) are exposed to policies and practices from other contexts. It is then trusted that this will lead to improvements in their own policies and practices and that they generally promote policy transfer and learning between the actors involved. At the same time, the reasons why a certain practice was considered worthwhile in a specific context may be deeply rooted in the social and political context. In the context of INTERREG B, 'good practice guides' usually consist of descriptive sequences of pilot projects, and many projects simply apply the term 'best/good practice' to all their case studies or pilot projects. Neither projects as the senders of 'good practice' nor their potential receivers usually address the fact that they are rather unique and difficult-to-transfer circumstances and evaluations of the policy innovation and experiments are seldom carried out (Wolman et al. 2004; Wolman and Page 2002). Instead, practices and ideas are simply described, and it is left to the potential recipient to decide which is the most appealing or has the highest validity in highly differentiated audiences (Wolman and Page 2002). Positive reputations of best practices then simply snowball as observers become self-referential. Moreover, using the dissemination of best practices as a means to get positive attention to one's own region or project means that only the 'good news' is circulated, while constructive contribution of negative lessons is ignored (Strang and Macy 2001). Successes do not usually invite reflection and may at best lay the basis for single-loop learning. Failures, on the other hand, are more often a source for learning, as they induce retrospective reflection and questioning of old habits, routines, and so on and are thus an important trigger for double-loop learning.

Besides, turning 'pilot projects', which are often meant for testing purposes and are unitary events, into 'good practice' contradicts the meaning of 'practice' as routine and repetition. On the contrary and depending on the type of transfer object, the transferability of one-off experiences to other settings, particularly between different countries, is highly limited (OECD 2001).

6.4 Examples of Transnational Project Approaches

In the following, five specific INTERREG IVB projects illustrate specific transnational working approaches. FUTURE CITIES, PORTICO, GREEN COOK and CCP21 made use of mutual feedback teams to reflect on pilot projects and potentially to improve these. MANDIE worked with a quite elaborate 'reality check' tool for its common conceptual basis.

A MANDIE: A Reality Check for a Transnational Concept for District Centre Management

MANDIE was concerned with the relatively new discipline of 'district centre management' and focused on district centres at the outskirts of cities. Project partners cooperated to stop the ongoing decline of these centres and to enhance their attractiveness and economic performance. They exchanged knowledge and experience to develop, implement and jointly evaluate innovative solutions. Codified results were training sessions, a lecture toolkit and a guidebook.

From a methodological perspective, the project was based on a transnational element: thematic working groups met to discuss and decide upon appropriate methods and strategies before pilot projects were implemented. A scientific advisory board ensured a comparative transnational analysis of conditions, characteristics, challenges and solutions. It also provided knowledge input and guidance to partners. However, in reality, the process did not take place as intended as project partners were more concerned with their individual pilot projects and needed time to familiarise themselves with transnational cooperation. A second transnational element included the joint evaluation of pilot projects and resulted in methodical guides for the determination of the generalisability and transferability of the developed methods and strategies. Finally, the subsequent application of methods and strategies in the pilot districts was supposed to further enhance their transferability with a 'reality check'.

In theory, this work approach made it possible to ensure a balance between transnational and local work as well as between conceptual, practical and evaluative phases in a process of several feedback loops. In practice, however, the joint concept was not the project's chronological starting point and could therefore only influence pilot projects to a limited degree. Also, the real-life application of the developed methods proved to be highly challenging.

B Transnational Feedback Groups in Practice

The objective of *FUTURE CITIES* was to enable city regions to cope with the predicted climate change impacts by transforming urban structures. To this end, project partners developed, applied and improved assessment criteria for 'climate proof cities'. In a first step, a consistent method was jointly developed to assess the climate proofness of cities with regard to strategic urban key components. Project partners then tested this tool in practice and developed local action plans and guidelines in parallel. The project-specific 'twinning approach' involved two partners from different countries who assessed and improved each other's action plans, taking into account best practice from at least two countries. These twinning partners were suggested on the basis of similar activities and challenges, and the hosting partner decided on the concrete problem to be discussed. Discussions mainly focused on very concrete, often technical and organisational aspects, but also on processes. These bilateral exchanges were described as very intense and fruitful. In contrast, discussions that touched upon more context-dependent topics (for example planning systems) were found to be less beneficial to partners. The exchange was concluded with a 'twinning report', which was not limited to individual insights but targeted towards the overall project. The approach was able to systematise and codify transnational exchange and its results. In essence, knowledge generalisation efforts were self-evaluated through practical application. Feedback loops improved the overall project results. While the project's first step, the joint development of an assessment method, was truly transnational, the twinning approach prevented partners from solely focusing on individual development and implementation.

The *PORTICO* partnership worked with the reconciliation of the limitations cities' cultural history poses to urban development with the opportunities cultural heritages provide. The project considered such restrictions a catalyst for innovation: an encouragement to develop new methods, techniques and ways of thinking about urban development. The working method of PORTICO was strongly based on the concept of 'transnational communities of practice' (a term originally coined by Lave and Wenger 1991). These were established for each work package to enhance processes of social and organisational learning by involving larger parts of partners' home organisations. They consisted of multi-disciplinary teams, with local and international experts connecting bottom-up and top-down expertise in three common themes. Local organisational representatives, experts from knowledge institutes and European knowledge networks and a mix of local decision-makers joined these 'laboratories'. They jointly developed the terms of reference for joint studies, visited all pilot projects and offered tailor-made advice and feedback to both pilot projects and the overall project.

A similar approach was pursued by the *GREEN COOK* project, which aimed at reducing food wastage and introducing a model for sustainable food management by changing the consumer-food relationship. For each work package, several 'communities of practice' brought together knowledge and know-how, set the terms of reference for action plans, methods and tools for various stakeholders, and tested, assessed and validating management models developed in pilot projects. Moreover, these groups developed and implemented dissemination and evaluation strategies, mobilised 'ambassadors' and networks and assessed the project's results to derive common standards. In this project, communities of practice only involved direct project partners. Additionally, the project organised a 'transnational coordination committee', where partners discussed the project's progress with an 'expert committee' that helped to solve practical problems, facilitated the implementation of actions and disseminated information to relevant networks.

CCP21 brought together seven major inland ports and promoted connectivity and sustainable transport by optimising the organisation of freight logistics and sustainable spatial development. Its working approach was based on 'cockpit teams' that were supposed to enhance the transnational learning process in each work package. These teams brought together expertise from project partners and external experts. They designed action plans, supervised their implementation, translated transnational learning goals into concrete outputs and organised the dissemination of results. With respect to pilot projects, each cockpit team delivered knowledge input and participated both in their implementation and evaluation.

All of these five projects included feedback groups comparable to those of the PARKS project (see section 5.3.4). However, instead of linking these groups to single pilot projects, they were often linked to work packages and thereby able to provide inter-pilot project feedback and exchange or simply served the overall project. They allowed for project structures that transcend pilot projects and emphasised transnational work. While GREEN COOK and FUTURE CITIES were limited to direct project partners, both PORTICO and CCP21 involved additional external knowledge input.

Developments in Transnational Methodologies: Conclusion

Classically, projects apply methods for experience accumulation such as study visits and pilot projects. With respect to knowledge codification, projects traditionally produce manuals, guidelines and toolboxes. However, their production can be individually different and is neither conceptualised by projects nor programmes. A mini-survey on the projects approved in the INTERREG IVB programme for Northwest Europe in March 2011 was able to identify two major types of transnational projects in Northwest Europe in terms of process phases: those focusing on the implementation of pilot projects and those with a logical starting point in the production of a joint analytical or conceptual basis for cooperation. With the latter, projects are able to reduce the challenges of transnational knowledge production that are caused by a portfolio of pilot projects of a low strategic fit. Pilot projects based on joint concepts have a common operational framework and permit the use of a common assessment method and thus enable joint sense making. This approach also provides pilot projects with a stronger transnational purpose.

Projects analysed in the survey also used mutual feedback, the evaluation of their pilot projects, the harmonisation of measures and early identification of potential knowledge 'senders' and 'receivers' to support knowledge development and learning. Feedback approaches for knowledge articulation such as those used in the PARKS project can be found in a variety of NWE projects. Processes of interactive and double-loop feedback, where the feedback receiver acknowledges the usefulness of feedback and avoids feedback of little use, partly happened in PARKS. Further research is needed to assess the experience with these transnational project elements and if and in how far they support transnational work and the production of transnational results.

Table 6.1 sums up the types of transnational working methods applied in the Northwest Europe INTERREG IVB programme at the time of the survey.

Table 6.1 Learning processes and transnational working methods found in INTERREG B NWE projects

Learning processes			
Level of analysis	*Experience accumulation*	*Knowledge articulation*	*Knowledge codification*
Individual	**Pilot project** Other tasks		**Learning journals** Reporting system (progress reports)
Group/ project	**Study visits**	**Transnational Feedback Sessions**	**Manual, guidelines, toolboxes**
	Staff exchange State-of-the-art research Presentations/ symposia	**Scientific Boards** **'Twinning approach', 'Communities of Practice' Transnational facilitators**	Meeting minutes

Source: adapted from Prencipe and Tell 2001, project applications NWE programme March 2011

7 Building a House on Stone
From Thorough Beginnings to Multiple Ends

Conceptualising and deconstructing the phenomenon of transnational knowledge development and learning permits the identification of some of the major challenges to transnational cooperation and causes that may prevent projects from making full use of their learning potential. This chapter summarises the findings of the analysis of knowledge development and learning in four case studies from the INTERREG B programme for Northwest Europe. These helped to work out relevant structural premises to transnational learning and inter-linkages between structures and processes. The learning process was conceptualised in several phases that are specific to transnational cooperation and help to analyse the internal functioning of transnational projects.

In the following, the general relevance of knowledge and learning processes for transnational projects is discussed in light of the empirical findings. This is followed by a summary of the analytical model for researching transnational learning and the identification of linkages between the different parameters. The model provides a starting point for future research on transnational learning that allows approaching the subject in a systematic and targeted way. At the level of projects, the analytical model highlights the relevance of strategic project design and process management. Based on the findings from the four case studies, recommendations to both transnational projects and programmes are provided with respect to a possible increase of project quality and strengthening of result-orientation with the help of a more conscious management of knowledge development and learning processes. Finally, relevant future research needs are discussed.

7.1 Transnational Knowledge Development and Learning

Transnational cooperation projects are of different types and orientation, ranging from research-oriented projects in programmes such as the EU's HORIZON 2020 over projects that develop new strategies and concepts to those implementing practical solutions on the ground in programmes focused on practitioners such as INTERREG, URBACT or LIFE+. What they all have in common is that those cooperating combine their different knowledge bases and jointly develop new knowledge that allows them to generate the projected project results (for example strategies, plans, technologies, methods and tools), which are often a reflection of the achieved knowledge gains. Knowledge development is at the heart of most innovation processes and essential to the implementation of innovative solutions. Thus, supporting both the ability to innovate and result-orientation (as increasingly demanded by the new INTERREG V generation programmes) of transnational projects means supporting their ability to generate the necessary knowledge by way of making best use of their transnational learning potential.

Previous studies have focused on the question of how transnational project results are implemented by the participating organisations and show that in many transnational projects, including INTERREG projects, a widespread practical implementation of lessons learned is far from reality. These studies conclude that the main challenge lies in the lack of organisational learning (Lähteenmäki-Smith and Dubois 2006; Böhme et al. 2003; Dühr and Nadin 2007). However, in this respect it has to be questioned if projects produce valuable and transferable results in the first place. The production and implementation of project results need to be seen in close interrelationship, with transnational findings being the result and their implementation being the desired effect of a project.

So far, INTERREG programme rhetoric has focused on the potential of learning *from* each other, which illustrates the relevance of knowledge transfer. The potential of learning *with* each other, which is characterised by cross-fertilisation processes and required for producing joint project results, is still not put into effect to its full extent.

As seen in the case studies, in the INTERREG context, transnational projects are not primarily perceived as knowledge development and learning processes and do not inevitably follow a logic that primarily focuses on knowledge and joint results. The motivation to participate in a transnational project is often characterised by the interest in drawing funding to the region and implementing local action. The dominance of pilot projects and focus on their success leads to working processes that are strongly concerned with local action. As illustrated by the case studies, the success of individual pilot projects is, however, not a guarantee for producing transnational results.

Still, cross-fertilisation processes and transnational results in their various knowledge forms are based on combining knowledge bases from different contexts and offer highly valuable input to regional development and a potentially high degree of innovation. Both successful local action and negative lessons provide useful insights into causal relationships and are thus valuable learning sources. However, to achieve the projects' inherent potential, considerable effort has to be made to overcome the challenges of transnational cooperation. Transnational learning requires that lessons learned are shared and processed at transnational level. As the outcomes of these processes are uncertain and depend on the interaction of many factors, participants tend to focus on the more direct benefits in the form of local action, which they are more likely to be able to control themselves. From the perspective of project partners, this mode of thought is understandable, also when taking into account the lack of incentives to produce project results that are of transnational and transferable character. The latter are mainly in the interest of programmes that aim at producing an added value that transcends individual action.

Project partners are not necessarily aware of the full potential benefit of transnational cooperation for their local project through knowledge transfer, observational learning and feedback. Nor are they necessarily aware of the role transnational knowledge and learning processes play for reaching projected project results or of the benefit that transnational results such as guidebooks, tools or strategies may have for their own local context. Knowledge and learning processes are not ends in themselves but means for the generation of transnational project results. They are complex, influenced by other processes (such as local political support or group socialisation) and run under the surface of projects. Naturally, they are thus not in the focus of project partners. The four case studies confirm how difficult it is to pay attention to knowledge development processes at the project level and consequently, project partners 'usually recognize the potential value [of learning], but it is not a focal point of the project, and they are unsure about how to proceed to harness the opportunity' (Sense 2003: 4). If at all, project teams have an underlying assumption that learning occurs randomly and uninhibitedly during the project (ibid.). In reality, however, the situation is more or less the opposite: 'learning within a project does not happen naturally; it is a complex process that needs to be managed. It requires deliberate attention, commitment, and continuous investment of resources' (Ayas 1998: 90).

Understanding the preconditions of transnational knowledge development and its process character is a prerequisite for identifying and actively facilitating learning processes in project teams. By supporting the flexibility and adaptability of organisations to the environmental challenges they are facing, transnational learning both helps to work more efficiently towards reaching project results and improving project quality and individual benefits from transnational cooperation.

At the same time, from the perspective of funding programmes, the production of lasting and transferable cooperation results that are not only of local benefit is the legitimation to fund transnational cooperation in the first place. The reconciliation of producing both direct local benefits and transnational results of potential benefit to the whole cooperation area and third parties is necessary if only conceptually highly challenging. The findings summarised in the following support the execution of transnational projects by enhancing the production of sufficient benefits for all sides and by overcoming some of the most demanding challenges to achieving transnational results.

7.2 An Analytical Model and its Application to Transnational Projects

The analytical model for researching transnational knowledge development and learning processes developed in section 4.3 brings together insights from theoretical contributions ranging from educational over organisational and inter-organisational studies to policy transfer literature. It conceptualises structural parameters relevant for learning in both inter-organisational and projects settings in three clusters: partnership-, knowledge- and project strategy-specific parameters. Previous studies in the INTERREG context considered a selected range of project structures and their influence on 'project success', but a systematic overview on relevant parameters has been missing. The analytical model then adds a process perspective to the structural grid that maps processes of exchange, direct and indirect learning, knowledge transfer, and knowledge production and which takes the specificities of transnational cooperation into account. These phases illustrate the potential of transnational cooperation and do not presuppose that every phase will be unlocked by every transnational project.

Existing structural analytical models for inter-organisational and project cooperation highlight the fact that all structural parameters are potentially interlinked. These inter-linkages, however, remain little researched and the lack of connection to process models for learning and knowledge creation prevents the linking of structural and procedural aspects and thus of potential links between project properties and the process flow. Moreover, it can be assumed that structural parameters are of different strengths in terms of their potential influence on learning process and outcomes, but again, this thought is not yet been much researched. As a first step towards an integrated view on project learning that takes project properties, processes and potential links within and between the two into account, the analytical model was applied to four transnational case studies. The application of the model to real-life cases allowed filling the theoretical model with empirically identified inter-linkages between the applied parameters on the one hand and between project structures, learning processes and learning outcomes on the other hand. On many occasions, inter-linkages identified in previous studies were confirmed, but additional connections were also found.

The case study analysis did not aim at making inferences to the overall population of transnational cooperation projects or to transnational INTERREG projects by, for example, stating how many projects encounter certain barriers or what the most popular solutions are. It thus did not aim at representativeness and did not attempt to measure the identified parameters in any way. Inferences were instead made to theory and challenges that are or can become relevant are portrayed. The analytical model does not presuppose that the identified inter-linkages surface in every transnational project and different case studies may lead to the identification of additional links. However, links that are of relevance in some cases may also be of relevance in other cases if they surface in a specific case depends on the relevant project design and processes. Cases, where they do not surface, may simply have found effective ways to deal with them. More successful projects may be weaker data sources, as many relevant challenges of transnational learning do not surface when processes run smoothly. Still, successful projects serve as examples of how challenges are avoided; while more challenged projects show what challenges exist and how these can be overcome.

Particularly through the challenges encountered, the cases revealed an extensive amount of information on the difficulties of transnational learning and their potential causes. Similarly, the achievements of the four cases point towards possible ways to overcome relevant barriers. The cross-case analysis made it possible to detect challenges in one case and to find relevant solutions in another. In addition, the approximation of the learning outcomes in the four case studies permitted a general appraisal of their learning processes as well as the identification of particularly influential structural and procedural parameters. The following cross-case analysis allowed the identification of commonalities and first patterns that provide clues to what aspects play a particular role for transnational learning.

The analytical model is a first step towards a systematic conceptualisation of transnational learning. The limitation to a certain set of parameters implies that findings are limited both in breadth and depth and do not represent a closing model of transnational learning. Still, the model illustrates the complexity of transnational cooperation and provides a starting point for future research on transnational learning and knowledge development that allows approaching the subject in a systematic and targeted way. To find out more about the exact functioning of single parameters, their relative strength and precise inter-linkages requires future research that goes into more depth of selected parameters. For future research, it may also be beneficial to link up the process perspective with the implementation of project results and find out more about how project processes may be conducive to building up 'learning boundaries' to organisational learning.

The case study approach permitted an understanding of the causal relationships in cooperation projects, their properties and cooperation, joint sense making and production of transnational results. Document analysis, interviews and participant observation made it possible to use data from different sources and to limit potential awareness and memory filters as far as possible.

If the above-discussed limitations of this work are embraced, its findings provide many valuable insights into possible inter-linkages between project structures and the processes of transnational knowledge development and learning. At the level of projects, the model highlights the relevance of strategic project design and process management. It supports the design of good projects by identifying project properties that influence transnational learning processes and supports process management by deconstructing the different relevant process phases.

The list of inter-related parameters in Table 7.1 allows a comprehensive theoretical understanding of learning and causal relationships in transnational projects. It amends relevant linkages identified in literature with those found in the four case studies.

Moreover, based on the findings from the case studies, three general approaches to knowledge development and learning can be identified (see Figure 7.1). In approach 1, project partners focus on their own activities while the potential of the partnership is less taken into account. This limits knowledge development and learning to intra-organisational and regional learning if – at all – partners succeed in transferring lessons learned to their home organisations and regions. As other research shows (for example Panteia et al. 2010; Wink 2010; Dühr and Nadin 2007; Böhme et al. 2003), the latter is highly challenging. In approach 2, partners realise the benefits from making use of the transnational pool of expertise and engage in transfer activities of existing experience and knowledge. Transnational

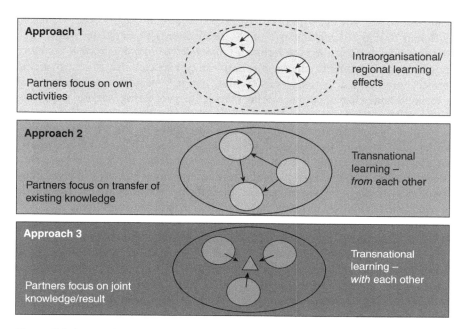

Figure 7.1 Approaches to knowledge development and learning in transnational projects

Source: by author

is possible by learning *from* each other. In approach 3, finally, partners jointly develop new knowledge by carrying out joint activities and tasks. Transnational learning is then possible by learning *with* each other. Taking into account that many projects are based on more than one knowledge process, these approaches can take place simultaneously with relation to different aspects.

7.3 Filling the Analytical Grid with Empirical Evidence from Transnational Cooperation

The analytical model for researching knowledge and learning processes in transnational cooperation projects consists of both parameters that describe given project properties and processes that develop over time. Based on inter-organisational and project-based learning literature, structural parameters are systemised in three clusters: partnership-, knowledge- and project strategy-relevant parameters. Project processes were identified based on organisational learning literature: (1) making new experience in pilot projects, (2) exchanging both new experience and existing knowledge, (3) reflecting on new experiences and lessons learned and (4) generalising transnational knowledge for the use of the overall project and to create transferable results. These contribute to the three knowledge functions of knowledge transfer, knowledge sharing, and knowledge development identified by Huelsmann et al. (2005). The process phases partly build on each other logically but do not, however, necessarily take place chronologically. They represent different degrees of making use of the transnational potential for knowledge development and learning: the further a project delves into the process, the more effort is required, but also the more benefit can be expected. Not all transnational projects are necessarily aware of how they can best benefit from transnational exchange and what importance joint knowledge processing has for producing truly joint results.

It quickly transpired that in many cases, the four projects were illustrations of the challenges of transnational knowledge development, but not necessarily of a successful handling of them, particularly with respect to knowledge generalisation. This lack of empirical findings is accompanied by a lack of substantial theoretical findings on the process of generalising from unique experiences in transnational settings. Thus, new insights into this matter remain limited.

Following the identification of process phases, the numerous inter-relationships between these and the structural parameters were worked out and summarised in Table 7.1. Some of the links were identified in theoretical contributions used and then confirmed by the cases studies (marked as 'L'), while others were found to play a role in the case studies, but have not yet been discussed in literature or only with respect to different dimensions (marked as 'C'). A note of caution has to be given: the inter-linkages identified in the table are not quantifiable, but indications of linkages. They do not make a statement on the representativeness of links but show the existing knowledge on the existence of inter-linkages. They require verification through in-depth studies. It is also possible that many more inter-linkages exist, which have not yet been identified in earlier studies and have not surfaced in the four case studies.

The table can be of help for project design by indicating potential implications of a certain choice of project properties (or lack of choice), it can be used for troubleshooting in projects by identifying potential barriers and their causal relationships and it can finally be of use for project assessment.

Many of the discussed parameters and processes are influenced by the context of INTERREG programmes. This means that while some of the findings are applicable to similar settings, others may be very specific to transnational cooperation in the INTERREG context. The latter include the inter-disciplinary character of the projects, their high share of practitioners (and thereby of tacit knowledge) and the practical implementation of pilot projects that influences the potential for 'observational learning', but also requires reflection processes.

In general, much points to a high impact of the complexity of the various parameters on all aspects of the knowledge development process. The four case studies reveal insights into many interesting inter-linkages and potential impacts. Among them, the different role that 'experts' and 'generalists' play for knowledge development and the different limitations they may face in terms of defining their 'knowledge roles' has not yet been discussed much. Moreover, the complexity and convertibility of 'knowledge roles' in a transnational cooperation project is an interesting aspect.

Table 7.1 Causal relationships between structural and process parameters in transnational knowledge development and learning

Identified causal relationships		Partnership-specific parameters					Strategy-specific parameters				Process parameters		
		Partner types	Experience and 'knowledge roles'	Motivation for cooperation	Composition of partnership	Knowledge	Objectives	Tasks and methods	Division of labour	Strategic fit of pilot projects	Exchange and communication	Knowledge transfer	Knowledge processing
Partnership-specific parameters	Partner types		C	L	L	L	L	C		C	L	L	L
	Previous experience and 'knowledge roles'	L	L	C	L	L	C	C	C	C	L	L	L
	Motivation for cooperation				L	L		C	C		L	L	L
	Composition of partnership			C	L	L	L	C		C	L	L	L
Knowledge-specific parameters	Knowledge types, characteristics	L	L	L	L		L	C		C	L	L	L
	Innovation phases				C	L	C		L		L	L	L
Strategy-specific parameters	Objectives	C		C	C	L		C	L	C	L	L	L
	Tasks and methods	C		C	C	C			C	C	L	C	L
	Division of labour									C	L	L	L
	Strategic fit pilot projects					C		C			C	C	C
Process parameters	Exchange and communication		C						C		L	L	L
	Knowledge transfer										L		L
	Reflection and feedback										C	L	L
	Knowledge systematisation										C	C	C
	Use of conceptual categories										L	L	L
	Knowledge abstraction											C	L

Source: by author

While the relevance of project objectives for project management is well known, the lack of methodical understanding and planning of transnational INTERREG projects is surprising. Against this background, the lack of guidance and a common direction in some of the case studies is understandable. Many transnational INTERREG programmes have grown sceptical towards the use of local investments. The arbitrariness of some of the pilot projects in the case studies and their detrimental impact on relevant knowledge development processes supports this scepticism. In the reality of INTERREG projects, pilot projects simply represent the implementation of the project at partners' localities. As they often receive the largest share of the funding and require notable resources, they are very often the central focus of project partners. However, without a direct functional role and relationship to each other and the overall project, they deprive projects of their transnational basis and turn them into an umbrella of individual sub-projects. On this basis, joint results are hard to produce and the future transferability to other contexts is limited.

A more effective use of the potential of pilot projects for hands-on learning requires better planning and project design. As seen, project design can lay strong foundations, but it is not a sufficient guarantee for reaching the best possible results. Conceptualising and supporting the various steps of the overall knowledge process can provide the required guidance to make the most use of transnational cooperation and to overcome its main barriers. Relevant, but little discussed in the INTERREG context, are the challenges connected to the time-related incompatibility of comprehensive knowledge input and the planning of project measures, to joint sense-making of experiential knowledge from local action and to the complexity of transnational knowledge generalisation on the basis of case-based experience. Many of these challenges are again linked to a general lack of a methodological conceptualisation of transnational cooperation among practitioners and point to the relevance of appropriate tasks and of a transnational methodology.

Table 7.2 summarises the theoretical contributions that were used to build the analytical model to show where case study evidence support theoretical findings. Aspects are identified that are worth exploring further with respect to processes of transnational learning in projects both from a theoretical and conceptual perspective and that have so far received only little attention in the context of transnational funding programmes. These can be of use to further conceptualise transnational cooperation processes and to create better conditions and results in transnational cooperation.

7.4 Geographical Aspects of Transnational Knowledge Transfer and Learning

So far, the fact that the transnational side of knowledge transfer and learning processes causes differing framework conditions, interpretations, usage of concepts and terms and thus difficulties for generalisation, has not been well reflected in literature. A few attempts have been made in the INTERREG context, but future

Table 7.2 Overview of findings and their linkages to theoretical contribution, their potential to further theoretic elaboration and development options in the INTERREG B programmes

Parameter	Theoretical contributions	Evidence of linkages from case studies	Starting points for further elaboration of theoretical contributions	Development options for INTERREG B programmes
Organisational types	Salk and Simonin 2008 (impact of organisational type and sector, governance form), cultural studies (for example Gullestrup 2006, Hofstede 2001)	Consequences for working styles, communicative culture, possession of knowledge	The role of public institutions in projects, inter-linkages between different institutional types and different methodological/ ideological approaches	Perception of project partners beyond institutional/thematic types'; role of practitioners working with context-dependent project topics versus researchers working at a more generalised level and how the latter may be able to help the former develop transferable knowledge
Previous experience and potential 'knowledge roles'	Tuomi 1999 (resource-perspective of knowledge), impact of experience on knowledge processes (Cohen and Levinthal, Powell et al. 1996, Szulanksi 1996), Bandura 1979 (observational learning), Wolman and Page 2002 (necessary to exchange information), Inkpen 2000 (knowledge accessibility)	The role of different experience bases for transfer options, relevance of potential 'senders' and 'receivers' for knowledge transfer	Different role of experts and generalists	Role of early identification of potential 'senders' and 'receivers' of knowledge
Learning objectives and motivation to cooperate	Motivation for learning (Osterloh and Frey 2000, Pérez-Nordtvedt et al. 2008, Szulanksi 1996), Dühr and Nadin (individual motivation relevant for cooperation), role for 'reciprocal learning capacity' (Lubatkin et al. 2001)	Role of motivation for 'reciprocal learning capacity', Role of transnational motivations and objectives for knowledge and learning processes		Extend the understanding of transnationality by the factor of motivation

(continued)

Table 7.2 (continued)

Parameter	Theoretical contributions	Evidence of linkages from case studies	Starting points for further elaboration of theoretical contributions	Development options for INTERREG B programmes
Composition of partnership	Role of similarities (Saxton 1997, Lubatkin et al. 2001, Jemison and Sitkin 1986, Colomb 2007, Dühr and Nadin 2007, Bachtler and Polverari 2007, Lähteenmäki-Smith and Dubois 2006, van Bueren et al. 2002), role of differences (Lubatkin et al. 2001, Noteboom 2000, Child and Faulkner 1998, Nonaka and Takeuchi 1995, Di Vicenzo and Mascia 2008), diversity increases complexity (Granrose and Oskamp 1997), diversity as a potential for innovation (Iles and Hayers 1997), impact of diversity (Hambrick et al. 1998), role of balance (Lubatkin et al. 2001, Saxton 1997, Böhme et al. 2003, Reagans and Zuckerman 2001), spatial structures (Böhme et al. 2003, Lähteenmäki-Smith and Dubois 2006), project topics (Böhme et al. 2003, Lähteenmäki-Smith and Dubois 2006, Colomb 2007, Dühr and Nadin 2007), motivations (Dühr and Nadin 2007), problem similarity (van Bueren et al. 2002)	Relevance of strategic fit of partners, pros and cons of similarities and differences, relevance of different types of diversity for cooperation, relevance for complexity and innovation potential		More comprehensive conceptualisations of a 'strategic fit' of partners; role of similarity of the problem or 'territorial evidence' (see section 7.5)
Knowledge characteristics	Knowledge types (Lubatkin et al. 2001), knowledge characteristics (Salk and Simonin 2008, Argote et al. 2003), innovation phases (Schoen et al. 2005, Dolowitz 2009)	Influence of knowledge complexity and tacitness on learning processes	Conceptualising different learning processes with respect to different development and innovation phases	Conceptualisation of 'exploitation' and 'exploration' strategies; conceptualisation of different development and innovation phases in projects and taking into account different requirements of more knowledge- or implementation-oriented projects

Project objectives	Role of clarity (Slevin and Pinto 1987, Turner 2009, Nicolas and Steyn 2008, Ayas and Zenuik 2001), transitional vs. action aspects (Lundin and Söderblom 1995), relevance of 'goal interdependence (Lubatkin et al. 2001)	Relevance of clear project objectives		Focus on transitional aspects and distinguish between objectives and actions
Tasks and methods	Role of tasks for action in project (Lundin and Söderblom 1995), disadvantage for knowledge processes through heterogeneous tasks (Zollo and Winter 2001), learning mechanisms (Prencipe and Tell 2001)	Disadvantage of unclear, little structured tasks for knowledge process, role for concerted transnational action	Further elaboration of transnational methods for producing joint results	Projects can create stronger 'roadmaps' when linking methods and tasks
Division of labour, integration of partners	Role for creating ties between partners (Di Vicenzo and Mascia 2008, Saxton 1997, Ayas and Zenuik 2001, Keller 1986), relevance for trust-building (Klein-Hitpass et al. 2006), relevance for building a 'common language' (Pérez-Nordtvedt et al. 2008), relevance for cooperation intensity and depth (Colomb 2007)	Relevance of close ties for information sharing and cooperation intensity	Potential links between close ties, communication, transferability and participatory approaches	Relevance of organisation of meetings to create strong ties, innovative meeting approaches
Strategic fit of pilot projects	Böhme et al. 2003 (decrease projects' transnational dimension)	Impact on transnational dimension	Requirements for horizontal and vertical fit, comparability and complementarity of project clusters in social sciences	Conceptualisation of strategic fit of pilot projects (vertical and horizontal) and how to best use them for the production of transnational results

(continued)

Table 7.2 (continued)

Parameter	Theoretical contributions	Evidence of linkages from case studies	Starting points for further elaboration of theoretical contributions	Development options for INTERREG B programmes
Transnational exchange and communication	'Space of social interaction' for knowledge sharing (Nonaka and Takeuchi 1995), relevance for group learning (Kissling-Näf and Knoepfel 1998), relevance for cross-linking existing knowledge (Lullies et al. 1993), role of communication for project management (Adenfeldt 2010, Pinto and Pinto 1990), role of communication for knowledge sharing (Nonaka and Takeuchi 1995), role of communication for learning processes (Gergen 1995, Lave and Wenger 1991), role of communication for exchange of tacit knowledge (Hansen et al. 1999), language barriers (Hambrick et al. 1998, Holden 2002), communication management (Knippschild 2008)	Role of intense exchange and mutually accessible knowledge for learning processes; relevance of cross-linking knowledge sources; role of intense communication for cooperation, knowledge development and learning (also tacit knowledge), language barriers in transnational cooperation	Inter-linkage between exchange and tasks/methods in projects, how to include dynamic knowledge aspects in group learning processes	Role of strong platforms to enhance communication and exchange beyond traditional partner meetings, creation of platforms for knowledge input, guidance to nurture a helpful discussion culture, relevance of clarifying terms
Exchange on pilot projects	Observational learning (Bandura 1979), role of reflection (Ayas and Zeniuk 2001, Raelin 2001, Mezirow 1991), experiential learning (Kolb 1984, Engeström 1999)	Conditions for 'observational learning', role of reflection for knowledge generalisation	Further development of the experiential learning approaches to include group processes and potentially diverse partnerships	Ensure sufficient reflection time for observational learning, nurture 'reflective practice' in projects and programmes, transnational feedback and for working out transfer options, build in dedicated tasks for pilot project assessment
Transnational knowledge transfer	Transfer objects (Dolowitz and Marsh 2000, Mamadouh et al. 2002, De Jong and Mamadouh 2002, OECD 2001), supporting conditions and barriers (van Bueren et al. 2002, Simonin 1999a, 1999b, Dolowitz and Marsh 2000, Szulanski 1996), transfer mechanisms (Mason and Leek 2008), relevance of de-contextualisation (Vettoretto 2007, Humpl 2004, Wolman and Page 2002), degrees of transfer (Rose 1993)	Role of knowledge characteristics for transferability, role of different 'transfer objects', conditions for transfer, role of de-contextualisation	Knowledge transfer in environments with more diverse partnerships, including different partner types	Sufficient reflection time, early identification of potential transfer options

Reflection and feedback	Fundamental to learning (Vinke-de Kruijf et al. 2013, Keen et al. 2005, Raelin 2001), reflective practice (Ayas and Zenuik 2001)	Ad-hoc and institutionalised feedback to support reflection, reflective practice easier in case of joint working, sufficient time and knowledge about framework situation at other locations, strategic pilot project portfolios, motivation to cooperate, different development stages, strong discussion culture	Obstacles related to complexity and context-dependency; potential and barriers of transnational feedback could be further elaborated and possible 'feedback strategies' identified	Sufficient reflection time; transnational feedback could be used more actively to enhance the performance of the overall project
Knowledge abstraction and generalisation	'Knowledge combination' and systematisation (Nonaka and Takeuchi 1995, Crossan et al. 1999), relevance of conceptual categories (De Jong and Edelenbos 2007), 'knowledge externalisation' (Nonaka and Takeuchi 1995), interpretation (Kolb 1984, Engeström 1995), role of de-contextualisation (Potter 2004, Hassink and Lagendijk 2001), role of learning tools (Ayas and Zenuik 2001), cluster evaluation (Sanders 1997), lack of interest in knowledge generalisation in projects (Mainemelis 2001, Grabher 2004b, Scarborough et al. 2004, Keegan and Turner 2001, Sahlin-Andersson 2002, Sydow et al. 2004)	Role of knowledge systematisation, role of participatory and communicative approaches to knowledge generalisation from a cluster of projects, lack of interest in knowledge generalisation in projects	Further elaboration of the role conceptual categories play for projects with diverse tasks and activities; in which projects these are easier to identify; what methods help to identify them; methods for de-contextualisation, methods for generalisation of different project and knowledge types, role of target-orientation for knowledge generalisation	Use of conceptual categories for work package design; stronger requirements for generalisation, methodological support for generalisation, emphasis on target-orientation and relevant links to knowledge generalisation, usage of final reports and other products for knowledge generalisation

1 The example of RIVERS illustrates how the incorporation of public authorities in the project enforced by the Joint Technical Secretariat led to two separated groups with little exchange, two very different and incompatible philosophies towards public participation and subsequently frustration and inability to produce joint results (for example methodic approaches).

Source: by author

research is necessary to elaborate the relevant implications. As a variety of authors argues, the *transnationality* of the input can be mirrored in the transnationality of the output (for example Lähteenmäki-Smith and Dubois 2006). Input, in this respect, is often related to partner structures and topics. In the context of partner structures, the case studies could not verify this assumption. Although the WOOD project strongly gained from its 'regional cooperation approach', mutual visits and the construction of common infrastructure that required spatial proximity, the structural conditions and cooperation processes of the three other projects suggest that the impact of other parameters was at least as strong. Findings from inter-organisational literature also point to the fact that knowledge tacitness and governance difficulties are more relevant criteria for learning processes than geographical locations (Lubatkin et al. 2001).

Still, it can be concluded that spatial proximity between partners can support project design with truly transnational properties because partners have much in common and possibly need to cooperate to solve their issues. In case of similar geographical conditions among partners but a lack of spatial proximity, strong commonalities may still be found, but the challenges of transnationality increase. In case of functional similarities and a lack of geographical similarities, these challenges grow even larger and transnational learning and the production of joint results can become quite complex. The highest degree of challenge is then faced by partnerships without geographical or functional similarities. These challenges are all related to partner origin and can be tackled with the help of a project design that is able to overcome the negative effects related to partner differences. Moreover, project processes can include transnational elements, which can potentially outbalance a lack of the transnationality of partner structures. Thus, although geographical relationships between partners may create favourable framework conditions for knowledge development and learning, during project processes many more factors came into play, which may have an even stronger impact.

This also holds with respect to the choice of topic: previous research showed that in the case of 'common issues', learning is often limited to the regions involved and has less of a transnational dimension (Böhme et al. 2003; Lähteenmäki-Smith and Dubois 2006). This leads to the assumption that transnational projects work better if they focus on 'transnational issues' (Colomb 2007; Dühr and Nadin 2007). The empirical analysis of the previous chapters suggests a more differentiated examination. It can be argued that if a topic is not transnational as such, mechanisms and working methods need to be so even more. On the other hand, the example of the RIVER project shows that even in cases of a 'transnational issue' and spatial proximity and similar geographical patterns (along a river), a project can still encounter severe barriers to knowledge development and learning. In this case, this was due to an abandoned transnational project strategy and methodology, a lack of motivation to share knowledge and knowledge characterisations that impeded the knowledge process, including a high degree of tacit knowledge. In contrast, in the case of the SEWAGE project, the lack of spatial proximity, spatial inter-connectedness and of a 'transnational issue' was balanced by a sound transnational strategy and division of labour, active participation by all partners, a strong focus on the joint objective, joint results, pilot projects of high strategic fit with the project objective and the relative ease of deriving common results in a natural science context.

These findings lead to the conclusion that transnationality is also strongly influenced by strategy-related variables, such as transnational objectives, tasks, methods and division of labour, which can – if unfavourable – disrupt the influence of favourable transnational partner structures and topics. Even more, it can be affected by partners' motivation to participate and thus their likeliness of making use of its full potential. The case studies showed that the transnationality of partner structures was no guarantee for a transnational output, but that in certain cases the transnationality of methods, strategies and learning processes was able to balance out a lack of transnationality of partner structures or topic. Although transnational partner structures and topics may ease cooperation, for projects with different partner structures and topics, this may mean that they need to invest more into their strategy and transnational working methods to achieve transnational results.

Moreover, the relationship between locational project structures and project outputs could mask a relation between the similarity of the problem and project outputs. The similarity of the problem in turn can also be increased in case of partners with functional, but non-geographical, relationships. Although locational and thematic project structures constitute an important premise for transnational cooperation, project results also highly depend on project processes, particularly those related to cooperation and knowledge processing.

In addition, the idea of a spatial context and spatial identity may be a stronger factor than the simple spatial proximity between project partners. Perceived functional relationships, enhanced through a certain spatial construction (for example Paasi 2001), may be of relevance for actors' motivation to cooperate, their (perceived) 'strategic fit' as well as the (perceived) congruence of framework conditions and thus an increased mutual relevance of experience and increased transferability of knowledge. As spatial construction and identity are dynamic, they may also be considered as the result of transnational learning processes (for example new spatial visions and images in the PARKS project). It is up to future research to test this.

7.5 Recommendations for Transnational Projects and Programmes

Although the transnational strand of the INTERREG programmes served as an exemplary framework for transnational projects, many of the findings of this analysis are relevant for similar initiatives, such as the EU's URBACT or LIFE+ programmes or programmes by the Nordic Council of Ministers. Helping projects to achieve their objectives in a more effective way is directly connected to knowledge and learning processes. Projects set out to produce new knowledge, to transfer knowledge, to apply knowledge and experience when implementing measures and to gain new knowledge from the experience of implementing them.

Projects that are target-oriented and both produce and apply new knowledge also contribute to target-oriented and effective programmes and ultimately to higher achievements with respect to the ERDF (or similar) objectives.

The findings in Table 7.1 are a step towards conceptualising knowledge processes in transnational cooperation. They do not make a statement on the efficiency of projects or their contribution to programme objectives and the objectives of the ERDF. The fact that some of the case studies met a considerable amount of barriers that (partly) prevented them from achieving results and learning processes of a high transnational benefit does not necessarily lead to the conclusion that the programmes that fund them are not working. INTERREG programmes are themselves part of a larger learning process if only one which is relatively slow and of little systematic nature. The transnational INTERREG programmes have been characterised by a learning-by-doing process (see section 2.4) and have considerably progressed throughout their funding periods by introducing stronger requirements for tangible results, transnational structures, innovation and strategy-orientation. As long as there are unused potential and room for improvement, the findings from the four case studies can help programmes and projects to avoid relevant barriers and to achieve projected results in a more effective way. Judging from the four case studies and the mini-survey, there are strong indications that transnational INTERREG programmes and projects do not yet realise the full extent of their knowledge production function. Possible causes that lead to low project performance with respect to achieving transnational objectives are identified in section 5.4

Many of the identified factors do not allow clear-cut and detailed recommendations, which require the consideration of their individual preconditions and of individual parameters in more depth. Observations from the case studies uncovered challenges that may not be easily solved, but nevertheless merit attention from project partners and designers and programme stakeholders. Still, some general recommendations can be derived.

In general, projects can gain from:

- consciously embracing knowledge development and learning as lying at the heart of transnational cooperation;
- identifying, constructing and managing learning processes;
- patience both with transnational cooperation in general and with the relevant knowledge processes;
- taking on responsibility for knowledge processes instead of leaving them up to coincidence (this trivial finding was fundamental in the case studies);
- awareness of the complexity of influencing factors and their inter-linkages on processes of transnational knowledge and learning.

The cross-case analysis of the four case studies identified seven structural aspects of particular importance for transnational learning and knowledge development. These influenced both other project properties and the learning processes in more than one case.

Partner Composition

1 *Create strategic partnerships*: While structural similarities between partners (for example professional background, institutional type) ease cooperation, from a certain extent, structural differences may become obstructive to cooperation (for example different regional challenges). In terms of knowledge, the situation seems reversed: differences in knowledge enhance options for knowledge transfer while too similar knowledge is of little attraction to transnational partners. In those cases, projects can still gain from cooperation, but their potential rather lies in joint knowledge development, which again requires an appropriate project strategy (objective, tasks, methods).

2 *Find partners with the right motivation for transnational cooperation*: Partner motivation to participate in a transnational project influences the extent to which partners are willing to integrate and participate as well as the overall process of knowledge development, that is the intensity of exchange, their participation in knowledge transfer and knowledge development exercises. Thus, attention needs to be paid to partner motivation and a strategic fit of partners.

Relevant Knowledge

3 *Manage knowledge complexity*: The complexity of knowledge crucially influences transnational learning. Knowledge complexity is again caused by the context-dependency of knowledge, limited relatedness due to highly diverse project tasks and pilot projects and limited options for verification. Knowledge complexity impacts on the whole process of transnational learning, including the usefulness of knowledge exchange, knowledge transferability and the development of joint knowledge by abstraction. It requires project strategies that embrace its challenges by formulating very clear objectives and projected results and applying appropriate work methods.

Project Strategies

4 *Draft clear project objectives and results*: The formulation of project objectives and projected results both strongly impact the conditions for transnational learning. Weakly formulated objectives and projects results may lead to a lack of partner motivation during the process. They also miss options for the reduction of knowledge complexity and joint sense making and negatively influence other strategy aspects, such as the definition of tasks, division of labour and methods. Moreover, they affect the entire learning process when knowledge exchange, transfer and development remain without clear goals. Therefore, project objectives should be formulated as clear and tangible as possible to allow the direct deduction of required tasks.

5 *Design transnational working methods*: The identification of transnational working methods supports cooperation, knowledge transfer and indirect learning from other pilot project experience as well as the development of

knowledge that combines project experience at all levels and lays the basis for transnational results. Therefore, projects need to reflect on how they will arrive at their joint results (including causal relationships) and gain from aligning methods for the overall project with those for individual pilot projects.

6   *Find a transnational division of labour*: The integration of project partners in the project tasks and the identification of a transnational division of labour that fosters task interdependence support both the motivation to actively engage in and thereby contribute to the overall learning process. Projects gain from a strong emphasis on cooperation in the sense of 'joint working' (joint tasks, actions and division of labour) that allow learning *from* each other, but also learning *with* each other and avoid fragmentation into individual pilot projects.

7   *Design a strategic pilot project portfolio*: The strategic fit of pilot project portfolios or a lack of it influences important knowledge characteristics, such as knowledge complexity and relatedness. Moreover, as observational learning and the transfer of knowledge from pilot projects depend on links between pilot projects, all learning process phases are affected. Thus, a strategic pilot project portfolio needs to be designed that ensures as much complementarity or comparability as possible as well as congruence with the overall project objectives and methods.

An additional parameter was identified that had not been part of the analytical model, but that nevertheless played a significant role in the case studies: the integration of relevant target group(s). Although the case studies do not show that the integration of target groups leads to better results, they illustrate how a lack of integration either complicated the production of project results or even contributed to the failure of producing joint results.

The findings in Table 7.1 allow a variety of recommendations to transnational projects, depending on the perspective one chooses to focus on. As already said, recommendations cannot always be of very clear-cut nature as the specificities of individual projects play a large role. This is particularly the case for the composition of partnerships, where for example partner heterogeneity leads to higher complexity that needs to be tackled, but that can also be beneficial in terms of creativity and the attractiveness of exchange. In case of 'common issues', the transnational strategy needs to be designed in a way that the disadvantages of a lack of 'transnational issues' are tackled and sufficient commonalities between partners are created (for example joint tasks, joint evaluation).

With respect to the *project strategy* and in addition to the recommendations provided above, more concrete recommendations can be given. Projects can gain in quality if they:

- pay attention to the design of their overall strategy (even if this is not required by the programme) as it influences the intensity of exchange and cooperation and options for knowledge processing;
- include a work package that synthesises the different project strands;

- work towards a clear 'product', which is based on the new knowledge gained in the different project components;
- gain an overview on the existing transnational knowledge pool at an early stage to enable timely identification of knowledge transfer options, an appropriate allocation of tasks in a way that they increase task interdependence and the optimisation of the pilot project portfolio with respect to its knowledge function;
- distinguish between tasks and the transitional aspect of a project when referring to project objectives;
- position the project with respect to its 'innovation phases'.

Recommendations for the *process management of knowledge development and learning* are challenged by the fact that they need to take the distinct project properties and conditions into account. Moreover, the analysis of the four case studies still partly leaves the question of how projects manage to overcome some of the challenges of transnational cooperation, as they, for example, disclosed the difficulties of transnational knowledge processing rather than solutions. However, some general recommendations on how to support knowledge development and learning processes can be provided:

- active involvement allows learning from others and access to the transnational pool of experience and knowledge;
- discussion and presentations at partner meetings that focus on common aspects increase the cooperative character of the project;
- active discussions are required for sharing, reflection and joint knowledge development;
- diversity management can help to handle some of the barriers of transnational cooperation particularly with respect to communication;
- openness about partner motivation and objectives helps to find commonalities to build upon;
- early discussions on the type and nature of possible conclusions support a targeted knowledge flow;
- the realisation of knowledge transfer options can be enhanced by identifying potential knowledge 'senders' and 'receivers' already at the start of the project;
- as the interpretation of information is not a neutral process and knowledge transfer requires the recipient side assesses the relevance and validity of the information for their own context, project partners gain from spending time to familiarise with the context of the transfer object and intense communication to access sufficient background information;
- a strategy that is able to deal with the barriers that transnational knowledge transfer faces in the context of the project topic supports the utilisation of transfer options, for example by taking into account different mentalities of target groups or by adapting knowledge to different contexts;
- transfer objects that are as concrete as possible, standardisation efforts, joint planning as well as a strategic pilot project portfolio, the coordinated assessment of experiences and joint communicative processes of sense-making and de-contextualisation support transnational knowledge transfer;

- the realisation of the full potential of pilot projects for transnational knowledge development requires sufficient exchange and reflection elements;
- new experience allows for additional learning, including already experienced partners and questioning of 'old ways', which potentially supports double-loop learning, but new insights need to be made accessible to project partners in order to be included in the joint results;
- be aware that reflection requires time and appropriate tools;
- generally embrace the complexity of transnational knowledge processing;
- provide a sound storing system for knowledge, capture and codify new insights and make it accessible to all partners;
- find cross-cutting issues and conceptual categories to which partners can relate and that support the synthesis of different knowledge strands;
- create options for abstraction and generalisation of knowledge; an inclusive and communicative approach ensures the identification of all partners with the results;
- produce tailor-made project outcomes and results for relevant target groups and include these during the project process where possible.

Many of the findings in Table 7.1 also have implications at the level of funding programmes, as these are responsible for setting the framework for project design and process. Although some of these are already considered when it comes to project selection at the INTERREG programme level, they are often applied in a very formalistic sense, only rudimentary and are so far not perceived in all relevant dimensions. Requirements introduced by the INTERREG programme level often seem to be initial responses to political demands for effective and result-oriented programmes, but do not always seem to be thoroughly considered. Partnership complexity may have been conceptualised in terms of location, geographical links and organisational types, but different levels of expertise or motivation and the dependence of strategic partnership on the project type and strategy play less of a role.

For transnational INTERREG programmes to create project conditions that support transnational knowledge development and learning and thereby the creation of transnational project results, it is beneficial to take into account the following points:

Partnership

1   Considering transnational projects in terms of the geography of the involved partners has been widely practiced, but this has often involved thinking in political terms rather than in functional, that is considering the participation of certain regions and countries on the basis of 'territorial evidence' (see box) and their usefulness for the cooperation process. Institutional types are generally considered, but not much allowance has yet been made for the concrete benefit that partners contribute. The new 2014 – 2020 INTERREG programmes now seem to propose a more systematic scan of the potential

contribution of partners' previous experience and knowledge in application forms and as assessment criteria. In how far this early detection of prior knowledge will be able to support transnational knowledge transfer and raise the quality of partnerships will have to be seen.

Basing transnational projects on 'territorial evidence'

Transnational partnerships are often created on the basis of a general topic, in which partners are interested while geographical or functional commonalities are not always considered despite their important role for ensuring beneficial cooperation. An alternative approach to assembling partnerships could be based on a strategic identification of those regions, organisations and expertise that are most relevant and particularly appropriate for the respective project objective. Territorial evidence could be used to argue why certain regions should cooperate and would open up the potential of stronger inter-linkages between the INTERREG programmes and the ESPON programme (European Spatial Planning Observation Network). Project partners could gain from setting their regional challenges in perspective to other European regions and as a tool for finding the most relevant partners.

Knowledge

2 In the INTERREG programme context, project knowledge is considered with respect to topics, either with regard to projects contributing to a certain programme priority or with regard to the differentiation into 'transnational issues' and 'common issues' (see section 2.5). The implications of certain knowledge types and characteristics for project strategies and transnational working methods, for example with respect to knowledge complexity and context-dependency and the relevance of accurate project objectives and results are – in contrast – little discussed. Moreover, a stronger conceptualisation of different project types with respect to innovation phases that may have different requirements with respect to the degree of detail in the application would be beneficial for programmes that aim at supporting innovation and would better meet the requirements of different project types. More flexibility for projects to react to upcoming opportunities would support their target-orientation. How the relevant knowledge types, the integrated and inter-disciplinary character of projects will change as a result of a stronger sectoral orientation of the new INTERREG programmes will have to be seen.[2]

---

2 Programme priorities of the 2014–2020 funding period had to be chosen from a list of potential thematic priorities provided by the Common Strategic Framework (Regulation (EU) 1303/2013). This does not any longer include the possibility of funding projects on integrated urban and regional development as such.

Project Strategies

3   In general, little attention has been paid to project strategies by the programme level. The formulation of precise objectives may have received certain attention in the assessment and approval of project applications and the formulation of precise results has moved into the focus of programmes, but objectives have so far rarely been considered with respect to their transition aspects (see section 5.2.3). The latter helps projects to focus on the change they attempt to achieve and to formulate tangible and achievable results. A considerable step in this direction is now being made by the recently approved 2014–2020 INTERREG programmes, which emphasise the importance of projects indicating the 'change' they are aiming at.

4   To work with work packages has been a requirement in INTERREG programmes since the 2007–2013 funding period. A lack of work packages led to the PARKS project suffering from a lack of overall structure and direction. However, in case of the RIVERS projects, they were not able to lead to an effective structure either. Reasons for this include work packages being too complex, highly unbalanced partner integration, a lack of particular work methods and of integrative aspects that can systematise and summarise findings. Thus, although work packages theoretically represent a very useful way to structure transnational projects, this does not happen automatically and requires a certain design. Again, the new 2014–2020 INTERREG programmes now seem to make important steps towards a stronger systematic structuring of work packages and require these to be directly linked to both project objectives and projected results.

5   The development of a transnational methodology has been widely neglected both by projects and programmes. Although every transnational project is faced with the challenge of integrating diverse knowledge pools and experience bases into processes of joint sense making on the basis of limited evidence (pilot projects), their experience with useful and useless methods is not assessed. Although INTERREG programmes have continuously raised demands for projects to produce tangible and transnational results, guidance and assistance on how this can be achieved is not provided.

Transnational methods

As the case studies illustrate, there was a lack of methodic conceptualisation of how to achieve transnational projects objectives. The review of 50 NWE projects showed that during the INTERREG IVB period, more projects included elements of joint evaluation and feedback as well as of a joint knowledge or conceptual basis. As this assessment was based on project applications, it remains to be seen how projects were actually able to make use of these methods and if they contributed to projects' ability to produce joint results. Interviews with the Lead Partners of two of the projects with more elaborate methodological approaches hint to practical difficulties. Although a large amount of projects attempts to produce transferable results in the

Although a large amount of projects attempts to produce transferable results in the form of 'guidelines', 'manuals' and similar products, projects do still not consider how and on what evidence base these will be produced. It can be argued that this allows for 'learning-by-doing' processes, but the reality of projects shows that these do not happen automatically. At the same time, the mass of completed transnational projects holds the communal experience with the challenges of producing joint results, but as these are not evaluated and interpreted at programme level, they cannot be used for future cooperation. If projected results are supposed to be more than lip service and realistically feasible, projects need clarity on appropriate transnational methods. Application forms or handbooks do not yet invite projects to think in these terms.

Methodic conceptualisation of the production of transferable results is strongly linked with the aspect of generalisation. The latter requires not only processing methods but also certain structural conditions, such as a 'strategic fit' of pilot projects. Generalisation and the production of common and transferable results are not ends in themselves, but support the results that are of use not only to the projects that produced them, but also to other actors and whole programmes (multiplier effect).

6   Pilot projects have become a standard element of transnational cooperation projects. They are often the main motivation for partners to participate and are the projects' response to the programmes' requirement for 'tangible results'. As a standard element, the necessity to include pilot projects in transnational INTERREG projects is no longer questioned. Nevertheless, there are two major challenges connected to pilot projects in transnational projects:

a   They represent the non-transnational element of transnational projects, are usually of little comparability and in many cases of limited complementarity. This questions their contribution to transnational project objectives and projected results.

b   They challenge project chronology when aiming at demonstrating project findings while being implemented in parallel to the knowledge production process. If actually meant for demonstrating, they need to follow knowledge production to be able to implement innovative findings. If meant for testing, they require systematic assessment and equivalent adjustments of concepts, strategies, and so on.

Projects gain from designing strategic pilot project portfolios that include sufficient linkages. Programmes gain from asking projects to define the specific contribution of individual pilot projects at the application stage. In the new 2014–2020 funding period, some INTERREG B programmes now seem to follow a more conscious approach to pilot projects, namely the North-West Europe programme that demands projects to develop pilot project portfolios based on a joint concept (Interreg North-West Europe 2015a). However, this does not seem to have found its way into the assessment criteria by which projects are selected. Joint concepts for pilot projects are also required in other programmes (for example Interreg Baltic Sea Region 2015), while others include specific quality requirements to individual pilot projects, but do not take the overall portfolio of pilot projects into account (Interreg CENTRAL EUROPE 2015).

Processes of Learning and Knowledge Development

7   Understanding processes of knowledge development and learning that support the production of transnational project results can be of major support to project management. Although many projects know about instruments to support the general project exchange, knowledge transfer processes also benefit from appropriate methods and instruments that support their preconditions (such as overviews on partner prior knowledge) or the processes themselves.

8   A specific potential for learning in transnational cooperation is provided by the option for indirect learning, so-called 'observational learning'. Again, this can be actively supported, for example, by the design of a strategic pilot project portfolio.

9   Projects require opportunities for reflection beyond individual pilot projects; at programme events even reflection beyond single projects (project clusters) support reflective practice.

10   Systematisation of project experience and knowledge is a project management task and supports the capturing of insights and reflection of individual experience. Knowledge abstraction and generalisation are highly complex tasks in transnational cooperation and require a substantial amount of guidance. The experience from successful examples helps to inform future processes.

11   A more critical stance to and a programme debate of the 'best practice' approach may be appropriate (see section 6.3).

All in all, the very recently approved INTERREG 2014–2020 programmes include a variety of new requirements for projects that increase the relevance for strategic and systematic learning and knowledge development in projects. Most notably, the new focus on result-orientation requires projects to make optimal use of transnational learning options and process new knowledge in a way that it contributes to innovative and transferable project results. Single programmes even set up specific tools to help projects achieve their objectives, such as detailed risk assessments (Interreg North-West Europe 2015) or compulsory mid-term reviews (Interreg CENTRAL EUROPE 2015). Being a necessary step towards the production of joint results and based on the overall project experience, transnational knowledge processing will be of even more importance than before. Some of the new programmes particularly emphasise the need for project outputs to be transferable to other settings, the CENTRAL EUROPE programme even makes knowledge transfer from the project to third parties compulsory (Interreg CENTRAL EUROPE 2015). Again, this increases the relevance of transferable project results and attaches more importance to abstraction and generalisation. Finally, the new programmes advise projects do evaluate the achievements of their pilot projects and/or the overall project proceedings. The new NWE Programme even demands that lessons learned from these evaluation activities can already be applied during the projects' lifetime (Interreg North-West Europe 2015a).

Still, some of the identified challenges of transnational cooperation cannot easily be solved. Knowledge development and learning processes in transnational projects entail strong dynamic aspects, which need to be reconciled with the project strategy. These include integrating learning processes at different levels and from different sources: from existing internal and external knowledge pools, from pilot projects, partner feedback and joint reflection. Although an initial project phase in which state-of-the-art knowledge is gathered holds crucial benefits for projects with respect to the development of new knowledge, substantial timing problems have to be faced. These are mainly due to programme requirements for the definition of project activities in advance, the time limitation of transnational projects and – in the INTERREG context – their strong orientation towards implementation-oriented action.

Conceptualising the innovation dimension of transnational projects

In general, the specific relation of transnational INTERREG projects to innovation can be described as fostering a creative milieu and allowing the development of new solutions by testing and improving. In order to let this specific potential unfold, however, projects need to be more than a bundle of local activities and not too much limited by a rigorous input and output control. Moreover, programmes need a stronger conceptualisation of how transnational INTERREG projects can contribute to innovation. At present, programmes favour projects in the later phases of their innovation processes, which are concerned with substantiating and implementing innovation, while less weight is given to explorative action. As both project types compete with each other for funding, the favouring of implementation aspects forces projects to either cover more or less the whole innovation process or focus on its later stages. The latter is indisputably less complex and demanding.

Explorative actions and creative adjustment processes are often labelled 'exchange of experience', a term that has been associated with a somewhat negative connotation in transnational INTERREG programmes. The reason for this lies in the history of the transnational INTERREG strand with many projects of the early programme periods being limited to a very general exchange of experience and not being able to produce joint and tangible results. Still today, many activities and processes required for innovation, such as brainstorming, status-quo analyses and preparatory activities for knowledge transfer are associated with not being target-oriented. As a result, projects that work with more open-ended processes and, for example, allow transnational feedback to lead to adjustments, are potentially disadvantaged.

The drawback of the strong focus on project results is a positive discrimination of projects that

- are in later phases in their innovation process;
- are less innovative in general;

(continued)

(continued)

- are based on established partnerships that know each other well and are possibly following a predecessor project;
- are of the type 'shopping list', where partners exactly know what they want to achieve individually, while not actually requiring transnational cooperation; in effect this means that options for transnational cooperation (such as transnational feedback and its implementation) are limited.

To deal with the lack of conceptualisation of projects as innovation processes is a challenge. The lack of tangible results and programmes' fixation on input factors (budgets, personnel, and so on) during the early programme periods in the 1990s explains why the outcome and result perspective – that demands detailed descriptions of planned results – is regarded as progressive. During the 2007–2013 funding period, the transnational INTERREG strand put particular emphasis on the formulation of tangible outputs and results to increase the target-orientation of projects. While defining specific objectives and projected results have many advantages, they can also form a tight corset for the project process that does not live up to the fact that many partners do not know each other well and that many options only develop and emerge over time. The requirement for defining specific activities and outputs thus limits creative project processes, which are 'dynamic, non-linear, and inductive process[es] of joint discovery that is contingent on behaviour, cognitive, and administrative factors, as well as luck' (Lubatkin et al. 2001: 1374). After all, programmes do not have the means to verify if envisaged project results are realistic at all.

The call for a stronger process perspective in transnational projects is inevitably linked to the question of project flexibility. Cooperation processes are often highly difficult to predict, a fact that is increased by transnationality and thematic complexity and requires a certain degree of flexibility. With stronger demands to describe projected project results at the application stage in detail, programmes have, however, been developing in the opposite direction. Flexible reaction to changing environments and changing perceptions of environments by mending and adjusting, on the other hand, allows projects to benefit from additional options and bandwagon effects. Thus, a lack of flexibility with respect to results may implicate disadvantages due to a lack of openness to the transnational process. This is particularly relevant for projects in early innovation phases as innovation is typically difficult to predict. However, flexibility may not be helpful in all respects. Unspecific objectives had a detrimental effect in the case studies, especially as they were not further clarified during the project process. Thus, flexibility at the application stage needs to be combined with equivalent clarification at a later stage.

Instead of only focusing on projected results, programmes could pursue a stronger internal conceptualisation of different project types, their needs and their specific 'niche' to innovation processes. This requires a stronger conceptualisation at programme level of what kinds of projects are beneficial for what kinds of political objectives and a general switch of attention from project topics to conceptual and methodic approaches.

Another dimension of transnational cooperation is the strategic dimension of partnership structures and governance levels, which can be of very different types and quality and project purposes, which can be of very diverse types suggests a look at transnational cooperation in the EU framework from the perspective of Europeanisation processes (for example Hachmann 2011; Dühr and Nadin 2007; Giannakourou 2005). Transnational project strategies range from open horizontal exchange processes over joint adoption processes of EU regulations to joint policy uploading processes. This invites a project typology according to horizontal, top-down and bottom-up Europeanisation processes. As this also concerns project internal knowledge development processes, it may add valuable perspectives on knowledge and learning processes. As an example, projects that work in a top-down direction are likely to be characterised by more concrete project objective and by *existing and de-contextualised concepts* (such as EU regulations). Working with de-contextualised concepts enhances familiarity with a high level of abstraction and increases the likelihood of experience from pilot projects being followed up by knowledge generalisation. This potentially strong effect on knowledge and learning processes is worth further research. Another example is the bottom-up approach in projects that intend to influence a governance level higher than their own with the help of joint strategies and concepts. Both the SEWAGE and PARKS project included elements of this 'policy-uploading' but did not incorporate the actual 'uploading' in the project strategy. Such projects are likely to have specific joint outputs and a concrete intended transition. Moreover, they potentially include a higher necessity for participatory approaches, equal and active knowledge input and constant feedback processes in order to produce joint policies, which again enhance knowledge development and learning processes. Research targeted towards analysing the different influences of this project typology on project processes could provide valuable additional insights into knowledge and learning in transnational projects that may be able to further promote the effectiveness of transnational cooperation.

7.6 Future Research Needs

The following list summarises the need for future research in areas related to transnational knowledge development and learning:

- the linking of a perspective on process with a perspective on implementation allows studying the organisational learning processes taking place and the detection of potential 'stickiness';
- the experience projects are making with more systematic approaches to transnational cooperation, for example by including a joint conceptualisation phase, systematic evaluation of experiences and subsequent adaptations to actions/concepts as well as 'reality checks' for joint project results;
- what different project types exist, what different requirements and roles for innovation these have and what political objectives these different projects can serve;

- what different types of pilot projects fit to different project types;
- the development of innovative concepts for partner identification, for example, based on 'territorial evidence';
- research into additional parameters, particularly with respect to processual parameters, such as group socialisation or contextual aspects (for example political environment);
- more knowledge is needed on how projects can effectively deal with the challenges of generalisation and abstraction to produce joint results that are also of relevance to actors beyond the immediate project scope. The concept of 'cluster evaluation' could be a useful starting point, and insights are required about relevant barriers and challenges and how these can be overcome.

Designing transnational ETC programmes has been characterised by discussions of the degree of implementation-orientation of projects. During the first programme period (1996–1999), very few projects worked towards the implementation of findings, which became one of the main critiques about the programmes. During the next phase (2000–2006), the implementation-orientation of transnational projects increased significantly, however, due to many very local investments at the expense of project transnationality. A requirement during 2007–2013 programme period was thus to include only those investments that are of transnational character. An analysis of how much the claims for more transnational results and structures from the side of the programme level in the ETC context were actually taken up by projects and how this impacts the quality of project results would be an interesting field of research. Similarly, how much the projects that will be approved during the 2014–2020 funding period will be able to work more result-oriented will be worth assessment.

Bibliography

Adenfelt, M. (2010): Exploring the performance of transnational projects: Shared knowledge, coordination and communication, *International Journal of Project Management*, vol. 28, pp. 529–38.

Albach, H. (1998): 'Kreatives Organisationslernen', in Albach, H., Dierkes, M., Berthoin Antal, A. and Vaillant, K. (eds): *Organisationslernen – Institutionelle und Kulturelle Dimension*, Berlin: edition sigma, pp. 55–77.

Argote, L., McEvily, B. and Reagans, R. (2003): 'Managing knowledge in organizations: An integrative framework and review of emerging themes', *Management Science*, vol. 49, pp. 571–82.

Argyris, C., Schön, D. (1978): *Organizational Learning: A Theory of Action Perspective.* Reading MA: Addison-Wesley.

Ayas, K. (1998): 'Learning through projects: Meeting the implementation challenge', in Lundin, R. and Midler, C. (eds): *Projects as arenas for renewal and learning processes*, Dordrecht: Kluwer Academic Publishers, pp. 89–98.

Ayas, K. and Zenuik, N. (2001): 'Project-based learning: Building communities of teflective practitioners', *Management Learning*, vol. 32, no. 1, pp. 61–79.

Bachtler, J. and Polverari, L. (2007): 'Delivering territorial cohesion: European cohesion policy and the European model of society', in Faludi, A. (ed.): *Territorial Cohesion and the European Model of Society*, Cambridge, MA: Lincoln Institute of Land Policy, pp. 105–28.

Baird, J., Plummer, R., Haug, C. and Huitema, D. (2014): 'Learning effects of interactive decision-making processes for climate change adaptation', *Global Environmental Change*, vol. 27, pp. 51–63.

Bakker, R. M., Cambré, B., Korlaar, L. and Raab, J. (2011): 'Managing the project learning paradox: A set-theoretic approach toward project knowledge transfer', *International Journal of Project Management*, vol. 29, no. 5, pp. 494–503.

Bakker, R. M. and Janowicz-Panjaitan, M. (2009): 'Time matters: the impact of "temporariness" on the functioning and performance of organizations', in Kenis, P., Janowicz-Panjaitan, M. and Cambré, B. (eds): *Temporary Organizations: Prevalence, Logic and Effectiveness*, Cheltenham/Northampton: Edward Elgar, pp. 121–54.

Baltic Sea Region Programme 2007–2013 (2012): Programme under European Territorial Co-operation Objective and European Neighbourhood and Partnership Instrument, Final approved version 3.0 as of January 2012 CCI No. 2007CB163PO020. http://eu.baltic.net/download.php?type=file&id=1631 (accessed on 22 June 2015).

Bandura, A. (1979): *Sozial-kognitive Lerntheorie*, Stuttgart: Klett-Cotta.

Bateson, G. (1973): *Steps to an Ecology of Mind: Collected Essays in Anthropology, Psychiatry, Evolution and Epistemology*, London: Paladin, Granada.

BBSR (2009): *Transnational Cooperation in Europe: The German INTERREG B Experience. TransCoop 2009*, Bonn: BBSR, Berichte Band 32.

Becker, M. and Præst Knudsen, M. (2006): *Intra and Inter-Organizational Knowledge Transfer Processes: Identifying the Missing Links*, DRUID Working Paper, vol. 06–32.

Bennett, C. J. and Howlett, M. (1992): 'The lessons of learning: Reconciling theories of policy learning and policy change', *Policy Sciences*, vol. 25, no. 3, pp. 275–94.

Benson, D. and Jordan, A. (2012): 'Policy transfer research: Still evolving, not yet through?' Political Studies Review, vol. 10, no. 3, pp. 333–8.

Berthoin Antal, A. (1998): 'Die dynamik der theoriebildungsprozesse zum organisationslernen', in Albach, H., Dierkes, M., Berthoin Antal, A. and Vaillant, K. (ed.): *Organisationslernen – Institutionelle und Kulturelle Dimension*, Berlin: edition sigma, pp. 31–52.

Böhme, K., Josserand, F., Haraldsson, P., Bachtler, J. and Polverari, L. (2003): *Transnational Nordic-Scottish Cooperation: Lessons for Policy and Practice*, Nordregio Working Paper 2003: 3, Stockholm: Nordregio Working Paper.

Böhme, K., Richardson, T., Dabinett, G. and Jensen, O. B. (2004): 'Values in a vacuum? Towards an integrated multilevel analysis of the governance of European space', *European Planning Studies*, vol. 12, no. 8, pp. 1175–88.

Brady, T. and Davies, A. (2004): 'Building project capabilities: From exploratory to exploitative learning', *Organization Studies*, vol. 25, no. 9, pp. 1601–21.

Bruce, A., Lyall, C., Tait, J. and Williams, R. (2004): 'Interdisciplinary integration in Europe: The case of the Fifth Framework programme', *Futures*, vol. 36, no. 4, pp. 457–70.

Bueren, E. van, Bougrain, F. and Knorr-Siedow, T. (2002): 'Sustainable neighbourhood rehabilitation in Europe: From simple toolbox to multilateral learning', in De Jong, M., Lalenis, K. and Mamadouh, V. D. (ed): *The Theory and Practice of Institutional Transplantation. Experiences with the Transfer of Policy Institutions: An Introduction to Institutional Transplantation*, Dordrecht: Kluwer Academic Publishers, pp. 263–80.

Bulkeley, H. (2006): 'Urban sustainability: Learning from best practice?', *Environment and Planning A*, vol. 38, no. 6, pp. 1029–44.

Byrith, C. (2009): 'Plenary Speech', Annual Conference INTERREG IVB North Sea Programme, 24 June 2009, Egmond an Zee.

Capello, R. (1999): 'Spatial transfer of knowledge in high technology milieux: Learning versus collective learning process', *Regional Studies*, vol. 33, no. 4, pp. 353–65.

Carlile, P. R. (2004): 'Transferring, translating, and transforming: An integrative framework for managing knowledge across boundaries', *Organization Science*, vol. 15, no. 5, pp. 555–68.

Carlile, P. R. (2002): 'A pragmatic view of knowledge and boundaries: Boundary objects in new product development', *Organization Science*, vol. 13, no. 4, pp. 442–55.

CEC (2010): *Communication from the Commission – Europe 2020. A strategy for smart, sustainable and inclusive growth*, Brussels, 3.3.2010 COM(2010) 2020 final.

CEC (2008): *EU Cohesion Policy 1988–2008: Investing in Europe's future*, Inforegio Panorama, no 26, Luxembourg: Office for Official Publications of the European Communities.

CEC (2004): *Guidelines for a Community Initiative Concerning Trans-European Cooperation Intended to Encourage Harmonious and Balanced Development of the European Territory INTERREG III*, 2004/C 226/02, Brussels, CEC.

CEC (2000): *Guidelines for a Community Initiative Concerning Trans-European Cooperation Intended to Encourage Harmonious and Balanced Development of the European Territory INTERREG III*, Brussels, CEC.

Chelimsky, E. (1997): 'The coming transformations in evaluation', in Chelimsky, E. and Shadish, W. R. (eds): *Evaluation for the 21st century*, Thousand Oaks: Sage, pp. 1–26.

Child, J. (2001): 'Learning Through Strategic Alliances', in Dierkes, M., Berthoin Antal, A., Child, J. and Nonaka, I. (eds): *Handbook of Organisational Learning and Knowledge*, Oxford: Oxford University Press, pp. 657–80.

Child, J. and Faulkner, D. (1998): *Strategies of Cooperation. Managing Alliances, Networks, and Joint Ventures*, New York: Oxford University Press.

Choo, C. W. (1998): *The Knowing Organization: How Organizations Use Information to Construct Meaning, Create Knowledge, and Make Decisions*, New York: Oxford University Press.

Cohen, W. M. and Levinthal, D. A. (1990): 'Absorptive capacity: A new perspective on learning and innovation', *Administrative Science Quarterly*, vol. 35, no. 1, pp. 128–52.

Colomb, C. (2007): 'The added value of transnational cooperation: Towards a new framework for evaluating learning and policy change', *Planning Practice and Research*, vol. 22, no. 3, pp. 347–72.

Cook, S. and Brown, J. (1999): 'Bridging epistemologies: The generative dance between organizational knowledge and organizational knowing', *Organization Science*, vol. 10, no. 4, pp. 381–400.

Crona, B. I. and Parker, J. N. (2012): 'Learning in support of governance: Theories, models, and a framework to assess how bridging organizations contribute to adaptive resource governance', *Ecology and scoiety*, vol. 17: 32.

Crossan, M., Lane, H. and White, R. (1999): 'An organizational learning framework: From intuition to institution', *Academy of Management Review*, vol. 24, pp. 522–37.

CSD – Committee on Spatial Development (1999): *European Spatial Development Perspective*, Luxembourg: Office for Official Publications of the European Communities.

Czarniawska, B. (1997): *Narrating the organization: Dramas of institutional identity*, Chicago/London: University of Chicago Press.

Dabinett, G. (2006): 'Transnational spatial planning: Insights from practices in the European Union', *Urban Policy and Research*, vol. 24, no. 2, pp. 283–90.

Dabinett, G. and Richardson, T. (2005): 'The Europeanisation of spatial strategy: Shaping regions and spatial justice through government ideas', *International Planning Studies*, vol. 10, no. 3/4, pp. 201–18.

De Jong, M. (2004): 'The pitfalls of Family resemblance: Why transferring planning institutions between "similar countries" is delicate business', *European Planning Studies*, vol. 12, no. 7, pp. 1055–68.

De Jong, M. and Edelenbos, J. (2007): 'An insider's look into policy transfer in transnational expert networks', *European Planning Studies*, vol. 15, no. 5, pp. 687–706.

De Jong, M., Lalenis, K. and Mamadouh, V. D. (1991): 'Learning from samples of one or fewer', *Organisation Science*, vol. 2, no. 1, pp. 1–13.

De Jong, M. and Mamadouh, V. (2002): 'Two contrasting perspectives on institutional transplantation', in De Jong, M., Lalenis, K. and Mamadouh, V. D. (eds): *The Theory and Practice of Institutional Transplantation: Experiences with the Transfer of Policy Institutions*, Dordrecht: Kluwer Academic Publishers, pp. 19–32.

De Jong, M., Mamadouh, V. and Lalenis, K. (2002): 'Drawing Lessons about Lesson Drawing', in De Jong, M., Lalenis, K. and Mamadouh, V. D. (eds): *The Theory and Practice of Institutional Transplantation: Experiences with the Transfer of Policy Institutions*, Dordrecht: Kluwer Academic Publishers, pp. 283–199.

Desprès, C., Brais, N. and Avellan, S. (2004): 'Collaborative planning for retrofitting suburbs: Transdisciplinarity and intersubjectivity in action', *Futures*, vol. 36, no. 4, pp. 471–86.

Dewey, J. (1938): *Experience and Education*, New York: Collier Books.

Di Vicenzo, F. and Mascia, D. (2008): 'Temporary organizations, social capital and performance: An empirical analayis', Paper presented at 24th EGOS Colloquium, Amsterdam, July 2008.

Diduck, A., Sinclair, A., Hostetler, G. and Fitzpatrick, P. (2012): 'Transformative learning theory, public involvement, and natural resource and environmental management', *Journal of Environmental Planning and Management*, vol. 55, no. 10, pp. 1311–30.

Dierkes, M. and Albach, H. (1998): 'Lernen über organisationslernen. Einführung, überblick und resümee', in Albach, H., Dierkes, M., Berthoin Antal, A. and Vaillant, K. (eds): *Organisationslernen – Institutionelle und Kulturelle Dimension*, Berlin: edition sigma, pp. 15–30.

Dierkes, M. and Marz, L. (1998): 'Leitbilder als Katalysatoren des Organisationslernens. Technikentwicklung als Anwendungsfeld', in Albach, H., Dierkes, M., Berthoin Antal, A. and Vaillant, K. (ed.): *Organisationslernen – Institutionelle und Kulturelle Dimension*, Berlin: edition sigma, pp. 373–97.

Dietrich, P., Lehtonen, M. H. and Lehtonen, P. (2007): 'A contextual model for researching temporary organisations', Paper submitted to the Nordic Academy of Management conference in Bergen 9–11.8.2007.

Dolowitz, D. (2009): 'Learning by observing', *Policy and Politics*, vol. 37, no. 3, pp. 317–34.

Dolowitz, D. and Marsh, D. (2000): 'Learning from abroad: The role of policy transfer in contemporary policy-making', *Governance*, vol. 13, no. 1, pp. 5–25.

Dühr, S., Colomb, C. and Nadin, V. (2010): *European Spatial Planning and Territorial Cooperation*, Abingdon: Routledge.

Dühr, S. and Nadin, V. (2007): 'Europeanization through transnational territorial cooperation? The case of INTERREG IIIB North-west Europe', *Planning Practice and Research*, vol. 22, no. 3, pp. 373–94.

Easterby-Smith, M. (1997): 'Disciplines of organizational learning: Contributions and critiques', *Human Relations*, vol. 50, no. 9, pp. 1085–1113.

Easterby-Smith, M. and Lyles, M. A. (2008): 'Introduction: Watersheds of organizational learning and knowledge management', in Easterby-Smith, M. and Lyles, M. A. (eds): *The Blackwell Handbook of Organizational Learning and Knowledge Management*, Oxford: Blackwell, pp. 1–15.

Easterby-Smith, M., Lyles, M. A. and Tsang, E. W. K. (2008): 'Inter-organizational knowledge transfer: Current themes and future prospects', *Journal of Management Studies*, vol. 45, no. 49, pp. 677–90.

Elkjaer, B. (2008): 'Social learning theory: Learning as participation in social processes', in Easterby-Smith, M. and Lyles, M. A. (eds): *The Blackwell Handbook of Organizational Learning and Knowledge Management*, Oxford: Blackwell, pp. 38–53.

Engeström, Y. (1999): 'Innovative learning in work teams: Analyzing cycles of knowledge creation in practice', in Engeström, Y., Miettinen, R. and Punamäki, R.-L. (eds): *Perspectives in Activity Theory*, Cambridge: Cambridge University Press, pp. 377–404.

European Council (2013): *Regulation (EU) No 1299/2013 of the European Parliament and of the Council of 17 December 2013 on specific provisions for the support from the European Regional Development Fund to the European territorial cooperation goal*, Luxembourg: Office Journal of the European Union L 347/259.

European Council (2013): *Regulation (EU) No 1301/2013 of the European Parliament and of the Council of 17 December 2013 on the European Regional Development Fund and on specific provisions concerning the Investment for growth and jobs goal and repealing Regulation (EC) No 1080/2006*, Luxembourg: Office Journal of the European Union L 347/289.

European Council (2013): *Regulation (EU) No 1303/2013 of the European Parliament and of the Council of 17 December 2013 laying down common provisions on the European Regional Development Fund, the European Social Fund, the Cohesion Fund, the European Agricultural Fund for Rural Development and the European Maritime and Fisheries Fund and laying down general provisions on the European Regional Development Fund, the European Social Fund, the Cohesion Fund and the European Maritime and Fisheries Fund and repealing Council Regulation (EC) No 1083/2006,* Luxembourg: Office Journal of the European Union L 347/320.

European Council (2006a): *Council Decision of 6 October 2006 on Community strategic guidelines on cohesion (2006/702/EC)*, Luxembourg: Official Journal of the European Union.

European Council (2006b): *Council Regulation (EC) No 1083/2006 of 11 July 2006 laying down general provisions on the European Regional Development Fund, the European Social Fund and the Cohesion Fund and repealing Regulation (EC) No 1260/1999,* Luxembourg: Official Journal of the European Union.

European Council (2001): *Presidency Conclusions of the Göteborg European Council, 15 and 16 June*, Göteborg: European Council, http://ec.europa.eu/smart-regulation/impact/background/docs/goteborg_concl_en.pdf (accessed on 24 June 2015).

European Council (2000): *Presidency Conclusions of the Lisbon European Council, 23 and 24 March*, Lisbon: European Council, http://www.europarl.europa.eu/summits/lis1_en.htm (accessed on 24 June 2015).

European Council (1999): *Council Regulation (EC) No 1260/1999 of 21 June 1999 laying down general provisions on the Structural Funds*, Luxembourg: Official Journal of the European Union L 161/1.

Interreg CENTRAL EUROPE (2015): *Interreg CENTRAL EUROPE Programme – Application Manual. Part B: What projects we are looking for*, http://www.interreg-central.eu/fileadmin/user_upload/Downloads/First_call/Part_B_final.pdf (accessed on 23 June 2015).

Interreg North-West Europe (2015): Interreg North-West Europe 2014–2020 Programme Manual, http://www.nweurope.eu/5b/documents/NWE-Programme-Manual-v1.1.pdf (accessed on 23 June 2015).

Interreg North-West Europe (2015): Cooperation Programme Interreg North-West Europe 2014–2020, http://www.nweurope.eu/5b/documents/NWE_Cooperation_Programme_FINAL.pdf (accessed on 23 June 2015).

Evans, M. (2004): 'Understanding policy transfer', in Evans, M. (ed.): *Policy Transfer in Global Perspective*, Aldershot: Ashgate, pp. 10–44.

Evans, M. and Davies, J. (1999): 'Understanding policy transfer: A multi-level, multi-disciplinary perspective', *Public Administration*, vol. 77, no. 2, pp. 361–85.

Finger, M. and Brand, S. B. (1999): 'The concept of the "learning organization" applied to the transformation of the public sector: Conceptual contributions for theory development', in Easterby-Smith, M., Burgoyne; J. and Araujo, L. (eds): *Organizational Learning and the Learning Organization: Developments in Theory and Practice*, London: Sage Publications, pp. 130–56.

Flyvbjerg, B. (2004): 'Five misunderstandings about case-study research', in Seale, C., Gobo, P., Gubrium, J. and Silverman, D. (eds): *Qualitative Research Practice*, London: Sage, pp. 420–34.

Friedman, V. J. (2001): 'The individual as agent of organizational learning', in Dierkes, M., Berthoin Antal, A., Child, J. and Nonaka, I. (eds): *Handbook of Organisational Learning and Knowledge*, Oxford: Oxford University Press, pp. 398–414.

Friedmann, J. (2005): 'Globalization and the emerging culture of planning', *Progress in Planning*, vol. 64, pp. 183–234.

Galison, P. (1997): *Image and Logic: A Material Culture of Microphysics*, Chicago: The University of Chicago Press.

Gergen, K. J. (1995): 'Social construction and the educational process', in Steffe, L. P. and Gale, J. (eds): *Constructivism in Education*. Hillsdale, NJ: Erlbaum, pp. 17–39.

Gherardi, S. (1999): 'Learning as problem-driven or learning in the cace of mystery', *Organization Studies*, vol. 20, no. 1, pp. 101–24.

Giannakourou, G. (2005): 'Transforming spatial planning policy in Mediterranean countries: Europeanization and domestic change', *European Planning Studies*, vol. 13, no. 2, pp. 319–31.

Grabher, G. (2004a): 'Learning in projects, remembering in networks? Communality, sociality, and connectivity in project ecologies', *European Urban and Regional Studies*, vol. 11, no. 2, pp. 99–119.

Grabher, G. (2004b): 'Temporary architectures of learning: Knowledge governance in project ecologies', *Organization Science Studies*, vol. 25, no. 9, pp. 1491–514.

Granrose, C. S. and Oskamp, S. (1997): *Cross-Cultural Work Groups*, Thousand Oaks: Sage.

Gredler, M. E. (2001): *Learning and Instruction: Theory into Practice*, Upper Saddle River: Prentice Hall.

Gullestrup, H. (2006): *Cultural Analysis: Towards Cross-Cultural Understanding*, Aalborg: Aalborg University Press.

Hachmann, V. (2011): 'From mutual learning to joint working: Europeanization processes in the INTERREG B programmes', *European Planning Studies*, vol. 19, no. 8, pp. 1537–55.

Hachmann, V. and Potter, P. (2007): 'Transnational and intercultural skills for the management of European networks', *The International Journal of Diversity in Organisations, Communities and Networks*, vol. 7, no. 1, pp. 187–94.

Hall, P. (1993): 'Policy paradigms, social learning and the state', *Comparative Politics*, vol. 25, no. 3, pp. 275–96.

Hambrick, D. C., Canney Davison, S., Snell, S. A. and Snow, C. C. (1998): 'When groups consist of multiple nationalities: Towards a new understanding of the implications', *Organization Studies*, vol. 19, no. 2, pp. 181–205.

Hamel, G. (1991): 'Competition for competence and inter-partner learning within international strategic alliances', *Strategic Management Journal*, vol. 67, no. 1, pp. 133–9.

Hansen, M. T., Nohria, N. and Tierney, T. (1999): 'What's your strategy for managing knowledge?', *Harvard Business Review*, vol. 2, pp. 106–16.

Hartley, J. and Allison, M. (2002): 'GOOD, BETTER, BEST? Inter-organizational learning in a network of local authorities', *Public Management Review*, vol. 4, no. 2, pp. 101–18.

Hassink, R. and Lagendijk, A. (2001): 'The dilemmas of interregional institutional learning', *Environment and Planning C: Government and Policy*, vol. 19, pp. 65–84.

Healey, P. (1992): 'Planning through debate: The communicative turn in planning theory', *The Town Planning Review*, vol. 63, no. 2, pp. 143–62.

Hedberg, B. (1981): 'How organisations learn and unlearn', in Nyström, P. C. and Starbuck, W. H. (eds): *Handbook of Organizational Design 1*, New York: Oxford University Press, pp. 3–27.

Hergenhahn, B. R. and Olson, M. H. (1997): *An Introduction to Theories of Learning*, Upper Saddle River, NJ: Prentice Hall.

Hofstede, G. (2001): *Culture's Consequences: Comparing Values, Behaviors, Institutions and Organizations Across Nations*, 2nd edition, Thousand Oaks: Sage Publications.

Holden, N. (2002): *Cross-Cultural Management: A Knowledge Management Perspective*, London: Prentice-Hall, Pearson Education.

Huebner, M. and Stellfeldt-Koch, C. (2009): *Impacts and Benefits of Transnational Projects (INTERREG III B)*, Federal Ministry of Transport, Building and Urban Affairs (BMVBS) / Federal Office for Building and Regional Planning (BBR), Forschungen 138.

Huelsmann, M., Lohmann, J. and Wycisk, C. (2005): 'The tole of inter-organizational learning and self-organizing systems in building a sustainable network culture', *International Journal of Knowledge, Culture and Change Management*, vol. 5, no. 2, pp. 21–30.

Humpl, B. (2004): *Transfer von Erfahrungen – Ein Beitrag zur Leistungssteigerung in Organisationen*, Wiesbaden: DUV.

Iles, P. and Hayers, P. K. (1997): 'Managing diversity in transnational project teams: A tentative model and case study', *Journal of Managerial Psychology*, vol. 12, no. 2, pp. 95–117.

Inkpen, A. (2000): 'A note on the dynamics of learning alliances: Competition, cooperation, and relative scope', *Strategic Management Journal*, vol. 21, pp. 775–9.

Inkpen, A. and Crossan, M. (1995): 'Believing is seeing: Joint-ventures and organization learning', *Journal of Management Studies*, vol. 32, no. 5, pp. 595–618.

Inkpen, A. and Dinur, A. (1998): 'Knowledge management processes and international joint ventures', *Organization Science*, vol. 9, pp. 454–68.

Innes, J. E. and Booher, D. E. (1999): 'Consensus building and complex adaptive systems: A framework for evaluating collaborative planning', *Journal of American Planning Associations*, vol. 65, no. 4, pp. 412–23.

Interreg Baltic Sea Region (2015): *Programme Manual for the period 2014 to 2020*, http://www.interreg-baltic.eu/fileadmin/user_upload/how-to-apply/1_call_step_1/1-3. Programme_Manual.pdf (accessed on 22 June 2015).

INTERREG IVC (2013): *Study on the Exchange of Experiences*, Lille: INTERREG IVC Joint Technical Programme Secretariat.

Jacoby, W. (2000): *Imitation and Politics: Redesigning Modern Germany*, Ithaca/London: Cornell University Press.

James, O. and Lodge, M. (2003): 'The limitations of 'policy transfer' and 'lesson drawing' for public policy research', *Political Studies Review*, vol. 1, pp. 179–93.

Janin Rivolin, U. and Faludi, A. (2005): 'The hidden face of European spatial planning: Innovations in governance', *European Planning Studies*, vol. 13, no. 2, pp. 195–215.

Janowicz-Panjaitan, M., Bakker, R. M. and Kenis, P. (2009): 'Research on temporary organizations: The state of the art and distinct approaches toward "temporariness"', in Kenis, P., Janowicz-Panjaitan, M. and Cambré, B. (eds): *Temporary Organizations: Prevalence, Logic and Effectiveness*, Cheltenham/Northampton: Edward Elgar, pp. 56–85.

Jemison, D. B. and Sitkin, S. B. (1986): 'Corporate acquisitions: A process perspective', *Academy of Management Review*, vol. 11, pp. 145–63.

Kazepov, Y: (2004): 'Cities of Europe: Changing contexts, local arrangements, and the challenge to social cohesion', in Kazepov, Y. (ed.): *Cities of Europe: Changing Contexts, Local Arrangements and the Challenge to Urban Cohesion*, London: Blackwell, pp. 3–42.

Keegan, A. and Turner, J. R. (2001): 'Quantity versus quality in project-based learning practices', *Management Learning*, vol. 31, no. 1, pp. 77–98.

Keen, M., Brown, V. A. and Dyball, R. (2005): 'Social learning: A new approach to environmental management', in Keen, M., Brown, V. A. and Dyball, R. (eds): *Social Learning in Environmental Management: Building a Sustainable Future*, London: Eartscan, pp. 3–21.

Keller, R. T. (1986): 'Predictors of the performance of project groups in R&D organizations', *Academy of Management Journal*, vol. 21, pp. 715–26.

Khanna, T., Gulati, R. and Nohria, N. (1998): 'The dynamics of learning alliances: Competition, cooperation, and relative scope', *Strategic Management Journal*, vol. 19, no. 3, pp. 193–210.

Kissling-Naef, I. and Knoepfel, P. (1998): 'Lernprozesse in öffentlichen Politiken', in Albach, H., Dierkes, M., Berthoin Antal, A. and Vaillant, K. (eds): *Organisationslernen – Institutionelle und Kulturelle Dimension*, Berlin: edition sigma, pp. 239–68.

Klein-Hitpass, K., Leibenath, M. and Knippschild, R. (2006): 'Vertrauen in grenzüberschreitenden Akteursnetzwerken. Erkenntnisse aus dem deutsch-polnisch-tschechischen Kooperationsprojekt ENLARGE-NET', *disP*, vol. 164, pp. 59–70.

Knight, L. (2002): 'Network learning: Exploring learning by interorganizational networks', *Human Relations*, vol. 55, no. 4, pp. 427–54.

Knill, C. (2005): 'Introduction: Cross-national policy convergence: Concepts, approaches and explanatory factors', *Journal of European Public Policy*, vol. 12, no. 5, pp. 764–74.

Knippschild, R. (2008): *Grenzüberschreitende Kooperation: Gestaltung und Management von Kooperationsprozessen in der Raumentwicklung im deutsch-polnisch-tschechischen Grenzraum*, Dresden: IÖR-Schriften Band 48.

Kolb, D. A. (1984): *Experiential Learning*, Englewood Cliffs, NJ: Prentice Hall.

Koskinen, K. U., Pihlanto, P. and Vanharanta, H. (2003): 'Tacit knowldege acquisition and sharing in a project work context', *International Journal of Project Management*, vol. 21, pp. 281–90.

Kroesen, O., de Jong, M. and Waaub, J. P. (2007): 'Cross-national transfer of policy models to developing countries: Epilogue', *Knowledge, Technology and Policy*, vol. 19, no. 4, pp. 137–42.

Lähteenmäki-Smith, K. and Dubois, A. (2006): *Collective learning through transnational co-operation: The case of Interreg IIIB*, Nordregio Working Paper 2006: 2 (Stockholm, Nordregio).

Lane, P. and Lubatkin, M. (1998): 'Relative absorptive capacity and interorganizational learning', *Strategic Management Journal*, vol. 19, pp. 461–77.

Lane, P., Salk, J. E. and Lyles, M. (2001): 'Absorptive capacity, learning and performance in international joint ventures', *Strategic Management Journal*, vol. 22, no. 12, pp. 1139–62.

Lave, J. and Wenger, E. (1991): *Situated Learning: Legitimate Peripheral Participation*, Cambridge: University of Cambridge Press.

Lewin, K. (1951): *Field Theory in Social Science*, New York: Harper.

[LRDP] Local and Regional Development Planning Ltd (2003): *Ex-Post Evaluation of the INTERREG II Community Initiative (1994–1999): Brief Report*, London: LRDP.

Lubatkin, M., Florin, J. and Lane, P. (2001): 'Learning together and apart: A model of reciprocal interfirm learning', *Human Relations*, vol. 54, no. 10, pp. 1353–82.

Lullies, V., Bollinger, H. and Weltz, F. (1993): *Wissenslogistik. Über den Betrieblichen Umgang mit Wissen bei Entwicklungsvorhaben*, Frankfurt: Campus.

Lundin, R. A. and Söderholm, A. (1995): 'A theory of the temporary organization', *Scandinavian Journal of Management*, vol. 11, no. 4, pp. 437–55.

Maaninen-Olsson, E., Wismén, M. and Carlsson, S. A. (2008): 'Permanent and temporary work practices: Knowledge integration and the meaning of boundary activities', *Knowledge Management Research and Practice*, vol. 6, pp. 260–73.

Macharzina, K., Oesterle, M.-J. and Brodel, D. (2001): 'Learning in multinationals', in Dierkes, M., Berthoin Antal, A., Child J. and Nonaka, I. (eds): *Handbook of Organisational Learning and Knowledge,* Oxford: Oxford University Press, pp. 631–56.

Mainemelis, C. (2001): 'When the muse takes it all: A model for experience of timelessness in organizations', *Academy of Management Review*, vol. 26, no. 4, pp. 548–65.

Mairate, A. (2006): 'The "added value" of European Union Cohesion policy,' *Regional Studies*, vol. 40, no. 2, pp. 167–77.

Makino, S. and Inkpen, A. C. (2008): 'Knowledge seeking FDI and learning across borders', in Dierkes, M., Berthoin Antal, A., Child J. and Nonaka, I. (eds): *Handbook of Organisational Learning and Knowledge*, Oxford: Oxford University Press, pp. 233–52.

Mamadouh, V., de Jong, M. and Lalenis, K. (2002): 'An introduction to institutional transplantation', in De Jong, M., Lalenis, K. and Mamadouh, V. D. (eds): *The Theory and Practice of Institutional Transplantation: Experiences with the Transfer of Policy Institutions*, Dordrecht: Kluwer Academic Publishers, pp. 1–16.

March, J. G. (1991): 'Exploration and exploitation in organizational learning', *Organization Science*, vol. 2, no. 1, pp. 71–87.

March, J. G., Sproull, L. G. and Tamuz, M. (1991): 'Learning from samples of one or fewer', *Organisation Science*, vol. 2, no. 1, pp. 1–13.

Mason, K. and Leek, S. (2008): 'Learning to build a supply network: An exploration of dynamic business models', *Journal of Management Studies*, vol. 45, pp. 759–84.

McCann, E. and Ward, K. (2010): 'Relationality/territoriality: Toward a conceptualization of cities in the world', *Geoform*, vol. 41, pp. 175–84.

McFarlane, C. (2006): 'Crossing borders: Development, learning and the North-South divide'. *Third World Quarterly*, vol. 27, no. 8, pp. 1413–37.

Merriam, S. B., Cafarella, R. S. and Baumgartner, L. M. (2007): *Learning in Adulthood: A Comprehensive Guide*, San Fransico: Jossey-Bass.

Mezirow, J. (1999): *Transformative Dimensions of Adult Learning*, San Francisco: Jossey-Bass.

van Mierlo, B. (2012): 'Convergent and divergent learning in photovoltaic pilot projects and subsequent niche development', *Sustainability: Science, Practice and Policy*, vol. 8, no. 2, pp. 4–18.

Mirwaldt, K., McMaster, I. and Bachtler. J. (2008): 'Reconsidering Cohesion Policy: The Contested Debate on Territorial Cohesion', *European Policy Research Paper*, vol. 66.

Mitchell, W. and Singh, K. (1996): 'Survival of business using collaborative relationships to commercialize complex goods', *Strategic Management Journal*, vol. 17, no. 3, pp. 169–95.

Mostert, E., Pahl-Wostl, C., Rees, Y., Searle, B., Tàbara, D. and Tippett, J. (2007): 'Social learning in European river-basin management: Barriers and fostering mechanisms from 10 river basins', *Ecology and Society*, vol. 12, no. 1, 19 (online).

Newig, J., Günther, D. P. and Pahl-Wostl, C. (2010): 'Synapses in the network: Learning in governance networks in the context of environmental management', *Ecology and Society*, vol. 15, no. 4, 24 (online).

Nicolas, J. M. and Steyn, H. (2008): *Project Management for Business, Engineering and Technology: Principles and Practice*. Elsevier: Oxford.

Nonaka, I. (1994): 'A dynamic theory of organizational knowledge creation', *Organization Science*, vol. 5, no. 1, pp. 14–37.

Nonaka, I. and Takeuchi, H. (1995): *The Knowledge-Creating Company: How Japanese Companies Create the Dynamics of Innovation*, Oxford: Oxford University Press.

Nonaka, I., Toyama, R. and Byosière, P. (2001). 'A theory of organizational knowledge creation: Understanding the dynamic process of creating knowledge', in: Dierkes, M., Berthoin Antal, A., Child J. and Nonaka, I. (eds): *Handbook of Organisational Learning and Knowledge*, Oxford: Oxford University Press, pp. 491–517.

Nooteboom, B. (2000): *Learning and Innovation in Organizations and Economies*, Oxford: Oxford University Press.

Northwest Europe INTERREG IV B Programme (2010): Project Handbook – Guidance Notes, http://www.nweurope.eu/index.php?act=page&page=funding&id=357 (accessed on 20 September 2014).

NWE Joint Technical Secretariat (2003): *Guidelines for Project Promotors*, Version 2003, Lille: NWE Joint Technical Secretariat.

OECD (2001): *Best Practice in Local Development*, Paris: OECD.

Osterloh, M. and Frey, B. (2000): 'Motivation, knowledge transfer and organizational forms', *Organization Science*, vol. 11, no. 5, pp. 538–50.

Paasi, A. (2001): 'Europe as a social process and discourses: Considerations of place, boundaries and identity', *European Urban and Regional Studies*, vol. 8, no. 1, pp. 7–28.

Packendorff, J. (1995): 'Inquiring into the temporary organization: New directions for project management research', *Scandinavian Journal of Management*, vol. 11, no. 4, pp. 319–33.

Panteia and Partners (2010a): *Ex-Post Evaluation of INTERREG 2000–2006 Initiative financed by the Regional Development Fund (ERDF). Task 5: In-depth analysis of a representative sample of programmes. Programme: Baltic Sea Region INTERREG III B Neighbourhood Programme*, Zoetermeer: Panteia, http://ec.europa.eu/regional_policy/sources/docgener/evaluation/expost2006/interreg_de.htm (accessed on 20 June 2015).

Panteia and Partners (2010b): *Ex-Post Evaluation of INTERREG 2000–2006 Initiative financed by the Regional Development Fund (ERDF). Task 5: In-depth analysis of a representative sample of programmes. Programme: INTERREG IIIB North West Europe*, Zoetermeer: Panteia, http://ec.europa.eu/regional_policy/sources/docgener/evaluation/pdf/expost2006/expo_interreg_north_west_europe.pdf (accessed on 20 June 2015).

Panteia and Partners (2010c): *INTERREG III Community Initiative (2000–2006) Ex-Post Evaluation (No. 2008.CE.16.0.AT.016). Final Report*, Zoetermeer: Panteia, http://ec.europa.eu/regional_policy/sources/docgener/evaluation/pdf/expost2006/interreg_final_report_23062010.pdf (accessed on 20 June 2015).

Pawlowsky, P. (1994): *Wissensmanagement in der Lernenden Organisation*, Habilitationsschrift Universität Paderborn.

Pedrazzini, L. (2005): 'Applying the ESDP through INTERREG IIIB: A southern perspective', *European Planning Studies*, vol. 13, no. 2, pp. 297–317.

Pérez-Nordtvedt, L., Kedia, B. L., Datta, D. K. and Rasheed, A. A. (2008): 'Effectiveness and efficiency of cross-border knowledge transfer: An empirical examination', *Journal of Management Studies*, vol. 45, no. 4, pp. 714–44.

Peterlin, M. and KreitmayerMcKenzie, J. (2007): 'The Europe of spatial planning in Slovenia, planning', *Practice and Research*, vol. 22, no. 3, pp. 455–61.

Peters, B. G. (1997): 'Policy transfer between governments; The case of administrative reforms', *West European Politics*, vol. 21, pp. 181–205.

Piaget, J (1970): *Genetic Epistemology*, New York.

Pierson, P. (2000): 'Increasing returns, path dependence, and the study of politics', *American Political Science Review*, vol. 94, no. 2, pp. 251–67.

Pinto, M. B. and Pinto, J. K. (1990): 'Project team communication and cross-functional cooperation in new program development', *Journal of Product Innovation Management*, vol. 7, no. 3, pp. 200–12.

Polanyi, M. (1967): *The Tacit Dimension*, New York: Anchor Books.

Potter, P. (2004): *Generalizing from the Unique: Bridging the Gap between Local Learning and Transnational Learning in European Urban Research and Evaluation*, KNi papers 01/04. Cologne.

Powell, W., Koput, K. and Smith-Doerr, L. (1996): 'Interorganizational collaboration and the locus of innovation: Networks of learning in biotechnology', *Administrative Science Quarterly*, vol. 41, pp. 116–45.

Prange, C., Probst, G. J. B. and Rüling, C.-C. (1996): 'Lernen zu kooperieren – Kooperieren, um zu lernen', *Zeitschrift Führung und Organisation*, vol. 1, pp. 10–16.

Prencipe, A. and Tell, F. (2001): 'Inter-project learning: Processes and outcomes of knowledge codification in project-based firms', *Research Policy*, vol. 30, pp. 1373–94.

Raelin, J. A. (2001): 'Public Reflection as the Basis of Learning', *Management Learning*, vol. 32, no. 1, pp. 11–30.

Reagans, R. and Zuckerman, E. (2001): 'Networks, Diversity and Productivity: The Social Capital of Corporate R&D Teams', *Organization Science*, vol. 12, no. 4, pp. 502–17.

Robertson, D. B. (1991): 'Political conflict and lesson drawing', *Journal of Public Policy*, vol. 11, no. 1, pp. 55–78.

Rose, R. (1993): *Lesson-drawing in Public Policy: a Guide to Learning across Time and Space*, Chatham: Chatham House Publishers.

Roy, A. (2011): 'Placing planning in the world: Transnationalism as practice and critique', *Journal of Planning, Education and Research*, vol. 31, no. 4, pp. 406–15.

Sahlin-Andersson, K. (1996): 'Imitating by editing success: The construction of organizational fields', in Czarniawska, B. and Sevón, G. (eds): *Translating Organizational Change*, Berlin: de Gruyter, pp. 69–92.

Salk, J. E. and Simonin, B. L. (2008): 'Beyond alliances: Towards a meta-theory of collaborative learning', in Easterby-Smith, M. and Lyles, M. A. (eds): *The Blackwell Handbook of Organisational Learning and Knowledge Management*, Oxford: Blackwell, pp. 253–77.

Sanders, J. R. (1997): 'Cluster evaluation', in Chelimsky, E. and Shadish, W. R. (eds): *Evaluation for the 21st Century*, Thousand Oaks: Sage, pp. 396–404.

Sanyal, B. (ed.) (2005): *Comparative Planning Cultures*, Routledge: London.

Savi, R. and Randma-Liiv, T. (2013): 'Policy transfer in new democracies: Challenges for public administration', in Carroll, P. and Common, R. (eds): *Policy Transfer and Learning in Public Policy and Management: International Contexts, Content and Development*, Routledge: Oxon, pp. 67–79.

Saxton, T. (1997): 'The effects of partner and relationship characteristics on alliance outcomes', *Academy of Management Journal*, vol. 40, no. 2, pp. 443–61.

Scarborough, H., Swan, J., Laurent, S., Bresnen, M., Edelman, L. and Newell, S. (2004): 'Project-based learning and the role of learning boundaries', *Organization Studies*, vol. 25, pp. 1579–1600.

Schindler, M. (2000): *Wissensmanagement in der Projektabwicklung*, Lohmar: Josef-Eul-Verlag. Reihe Wirtschaftsinformatik Band 32.

Schoen, J., Mason, T. W., Kline, W. A. and Bunch, R. M. (2005): 'The innovation cycle: A new model and case study for the invention to innovation process', *Engineering Management Journal*, vol. 17, no 3, pp. 3–10.

Schusler, T. M., Decker, D. J. and Pfeffer, M. J. (2003): 'Social learning for collaborative natural resource management', *Society and Natural Resources*, vol. 16, no. 4, pp. 309–26.

Schüppel, J. (1997): *Wissensmanagement*, Wiesbaden: Deutscher Universitäts-Verlag.

Senge, P. (1990): 'The leader's new work: Building learning organisations', *Sloan Management Review*, vol. Fall, pp. 7–23.

Senge, P. (1990): *The Fifth Discipline: The Art and Practice of the Learning Organisation*, New York: Doubleday.

Sense, A. J. (2003): 'Learning generators: Project teams re-conceptualised', *Project Management Journal*, vol. 34, no. 3, pp. 4–12.

Simonin, B. (1999a): 'Ambiguity and the process of knowledge transfer in strategic alliances', *Strategic Management Journal*, vol. 20, pp. 595–623.

Simonin, B. (1999b): 'Transfer of marketing know-how in international strategic alliances: An empirical investigation of the role and antecedents of knowledge ambiguity', *Journal of International Business Studies*, vol. 30, no. 3, pp. 463–90.

Slevin, D. P. and Pinto, J. K. (1987): 'Balancing strategy and tactics in project implementation', *Sloan Management Review*, vol. 9, pp. 33–41.

Söderholm, A. (1991): *Organiseringens Logik: En Studie av Kommunal Näringslivspolitik*, Umeå University: Dept of Business Administration.

Sol, J., Beers, P. J. and Wals, A. E. (2013): 'Social learning in regional innovation networks: Trust, committment and reframing as emergent properties of interaction', *Journal of Cleaner Production*, vol. 49, pp. 35–43.

Stead, D., de Jong, M. and Reinholde, I. (2008): 'Urban transport policy transfer in Central and Eastern Europe', *disP – The Planning Review*, vol. 44, no 1, pp. 62–73.

Stone, D. (2004): 'Transfer agents and global networks in the "transnationalization" of policy', *Journal of European Public Policy*, vol. 11, no. 3, pp. 545–66.

Stone, D. (2000): 'Non-governmental policy transfer: The strategies of independent policy institutes', *Governance*, vol. 13, pp. 45–70.

Stone, D. (1999): 'Learning lessons and transferring policy across time, space and disciplines', *Politics*, vol. 19, no. 1, pp. 51–9.

Strang, D. and Macy, M. W. (2001): 'In search of excellence: Fads, success stories and adaptive emulation', *American Journal of Sociology*, vol. 107, no. 1, pp. 147–82.

Swan, J. (2003): 'Knowledge management in action?', in Holsapple, C. W. (ed.): *Handbook on Knowledge Management 1: Knowledge Matters*, Berlin: Springer-Verlag, pp. 271–96.

Sydow, J., Lindkvist, L. and DeFilippi, R. (2004): 'Project-based organizations, embeddedness and repositories of knowledge: Editorial', *Organization Studies*, vol. 25, no. 9, pp. 1475–89.

Szulanski, G. (1996): 'Exploring internal stickiness: Impediments to the transfer of best practice within the firm', *Strategic Management Journal*, vol. 17 (Winter Special Issue), pp. 27–43.

Szulanski, G, and Capetta, R. (2008): 'Stickiness: Conceptualizing, measuring, and predicting difficulties in the transfer of knowledge within organizations', in Easterby-Smith, M. and Lyles, M. A. (eds): *The Blackwell Handbook of Organizational Learning and Knowledge Management*, Oxford: Blackwell, pp. 513–34.

Thompson Klein, J. (2004): 'Prospects for transdisciplinarity', *Futures*, vol. 36, pp. 515–26.

Tsoukas, H. (2008): 'Do we really understand tacit knowledge?', in Easterby-Smith, M. and Lyles, M. A. (eds): *The Blackwell Handbook of Organizational Learning and Knowledge Management*, Oxford: Blackwell, pp. 410–27.

Tuomi, I. (1999): *Corporate Knowledge: Theory and Practice of Intelligent Organizations*, Helsinki: Metaxis.

Turner, J. R. (2009): *The Handbook of Project-based Management: Leading Strategic Change in Organizations*, New York: McGraw-Hill.

URBACT (2004): *Guide to Capitalization*, Brussels: Urbact.

Valkering, P., Beumer, C., de Kraker, J. and Ruelle, C. (2013): 'An analysis of learning interactions in a crossborder network for sustainable urban neighbourhood development', *Journal of Cleaner Production*, vol. 49, pp. 85–94.

Vera, D. and Crossan, M. (2008): 'Organizational learning and knowledge management: Toward an integrative framework', in Easterby-Smith, M. and Lyles, M. A. (eds): *Handbook of Organizational Learning and Knowledge Management*, Oxford: Blackwell, pp. 122–42.

Verres, R. (1979): 'Vorwort zur deutschen Ausgabe', in Verres, R. (ed.): *Sozial-kognitive Lerntheorie*, Stuttgart: Klett-Cotta, Reihe: Konzepte der Humanwissenschaften.

Vettoretto, L. (2009): 'A preliminary critique of the best and good practice approach in European spatial planning and policy-making', *European Planning Studies*, vol. 17, no. 7, pp. 1067–83.

Vinke-de Kruijf, J., Hulscher, S. J. M. H. and Bressers, J. T. A. (2013): 'Knowledge transfer in international cooperation projects: Experiences from a Dutch-Romanian project', in Chavoshian, A- and Takeuchi, K. (eds): *Floods from Risk to Opportunity*, Wallingford: IAHS publ., pp. 423–34.

Vreugdenhil, H., Slinger, J., Thissen, W. and Rault, P. (2010): 'Pilot projects in water management', *Ecology and Society*, vol. 15, no. 3, 13 online.

Vygotsky, L. S. (1978): *Mind in Society: The Development of Higher Psychological Processes*, Cambridge, MA: Harvard University Press.

Waterhout, B. and Stead, D. (2007): 'Mixed messages: How the ESDP's concepts have been applied in INTERREG IIIB programmes, priorities and projects', *Planning Practice and Research*, vol. 22, no. 3, pp. 395–415.

Weick, K. E. and Roberts, K. H. (1993): 'Collective mind in organizations: Heedful inter-relating on flight decks', *Administrative Science Quarterly*, vol. 38, no. 3, pp. 357–81.

Wink, R. (2010): 'Transregional institutional learning in Europe: Prerequisites, actors and limitations', *Regional Studies*, vol. 44, no. 4, pp. 499–511.

Wolman, H., Hill, E. W. and Furdell, K. (2004): 'Evaluating the success of urban success stories: Is reputation a guide to best practice?', *Housing Policy Debate*, vol. 15, no. 4, pp. 965–97.

Wolman, H. and Page, E. (2002): 'Policy transfer among local governments: An information-theory approach', *Governance*, vol. 15, no. 4, pp. 477–501.

Yin, R. K. (2003): *Case Study Research: Design and Methods*, Applied Social Research Methods Series. Thousand Oaks: Sage.

Zaucha, J. and Szydarowski, W. (2005): 'Transnational co-operation and its contribution to spatial development and EU enlargement: The case of INTERREG IIIB in Northern Poland', *Informationen zur Raumentwicklung*, vol. 11/12, pp. 731–40.

Zollo, M. and Winter, S. (2001): 'Deliberate learning and the evolution of dynamic capabilities', *Organization Science*, vol. 13, no. 3, pp. 339–51.

Index